THEORY OF MACHINES & MECHANISMS

AUTHOR

Prof. G. C. Mohan Kumar

Professor (HAG), Dept. of Mechanical Engineering
National Institute of Technology Karnataka, Surathkal
Srinivasnagar 575025, Karnataka- India

REVIEWER

Dr. Bhushan Dattatray Nandre

Associate Professor, Dept. of Mechanical Engineering
Modern Education Society's Wadia College of Engineering,
Pune, Maharashtra, India

KHANNA BOOK PUBLISHING CO. (P) Ltd.
4C/4344, Ansari Road, Darya Ganj,
New Delhi, India

All India Council for Technical Education
Nelson Mandela Marg, Vasant Kunj,
New Delhi, India

Dear Readers,

To prevent the piracy, this book is secured with HIGH SECURITY HOLOGRAM on the front title cover. In case you don't find the hologram on the front cover title, please write us to at contact@khannabooks.com or whatsapp us at +91-99109 09320 and avail special gift voucher for yourself.

Specimen of Hologram on front Cover title:

Moreover, there is a SPECIAL DISCOUNT COUPON for you with EVERY HOLOGRAM.

How to avail this SPECIAL DISCOUNT:

Step 1: Scratch the hologram

Step 2: Under the scratch area, your "coupon code" is available

Step 3: Logon to www.khannabooks.com

Step 4: Use your "coupon code" in the shopping cart and get your copy at a special discount

Step 5: Enjoy your reading!

ISBN: 978-93-5538-974-9

Book Code: DIP499EN

Theory Of Machines & Mechanisms

by Prof. G. C. Mohan Kumar

First Edition: 2026

Originally Published by:

All India Council for Technical Education, Ministry of Education, GOI

Exclusively Published & Distributed by:

Khanna Book Publishing Co. (P) Ltd.
4C/4344, Ansari Road, Daryaganj,
New Delhi (INDIA)

Visit us at: **www.khannabooks.com**
Write us at: **contact@khannabooks.com**

To view complete list of books,
Please scan the QR Code:

KPH

Printed in Bharat.

प्रो. टी. जी. सीताराम
अध्यक्ष
Prof. T. G. Sitharam
Chairman

सत्यमेव जयते

आज़ादी का अमृत महोत्सव

अखिल भारतीय तकनीकी शिक्षा परिषद्
(भारत सरकार का एक सांविधिक निकाय)
(शिक्षा मंत्रालय, भारत सरकार)
नेल्सन मंडेला मार्ग, वसंत कुंज, नई दिल्ली–110070
दूरभाष : 011–26131498
ई–मेल : chairman@aicte-india.org

ALL INDIA COUNCIL FOR TECHNICAL EDUCATION
(A STATUTORY BODY OF THE GOVT. OF INDIA)
(Ministry of Education, Govt. of India)
Nelson Mandela Marg, Vasant Kunj, New Delhi-110070
Phone : 011-26131498
E-mail : chairman@aicte-india.org

FOREWORD

Engineers are the backbone of any modern society. They are the ones responsible for the marvels as well as the improved quality of life across the world. Engineers have driven humanity towards greater heights in a more evolved and unprecedented manner.

The All India Council for Technical Education (AICTE), have spared no efforts towards the strengthening of the technical education in the country. AICTE is always committed towards promoting quality Technical Education to make India a modern developed nation emphasizing on the overall welfare of mankind.

An array of initiatives has been taken by AICTE in last decade which have been accelerated now by the National Education Policy (NEP) 2020. The implementation of NEP under the visionary leadership of Hon'ble Prime Minister of India envisages the provision for education in regional languages to all, thereby ensuring that every graduate becomes competent enough and is in a position to contribute towards the national growth and development through innovation & entrepreneurship.

One of the spheres where AICTE had been relentlessly working since past couple of years is providing high quality original technical contents at Under Graduate & Diploma level prepared and translated by eminent educators in various Indian languages to its aspirants. For students pursuing 2nd year of their Engineering education, AICTE has identified 87 books, which shall be translated into 12 Indian languages - Hindi, Tamil, Gujarati, Odia, Bengali, Kannada, Urdu, Punjabi, Telugu, Marathi, Assamese & Malayalam. In addition to the English medium, books in different Indian Languages are going to support the students to understand the concepts in their respective mother tongue.

On behalf of AICTE, I express sincere gratitude to all distinguished authors, reviewers and translators from the renowned institutions of high repute for their admirable contribution in a record span of time.

AICTE is confident that these outcomes based original contents shall help aspirants to master the subject with comprehension and greater ease.

(Prof. T. G. Sitharam)

ACKNOWLEDGEMENT

The author is grateful to the authorities of AICTE, particularly Prof. (Dr.) T G Sitharam, Chairman; Dr. Abhay Jere, Vice-Chairman; Prof. Rajive Kumar, Member-Secretary; Dr. Ramesh Unnikrishnan, Advisor-II and Dr. Sunil Luthra, Director, Training and Learning Bureau for their planning to publish the books on Theory of Machines and Mechanisms.

I would like to thank the staff, CAD/CAM laboratory, and my PhD scholars at the Mechanical Engineering Department, National Institute of Technology, Karnataka, for their valuable support in completing this book. I sincerely thank my friends, colleagues, and wife, Mrs Meenakshi, for their patience while I wrote this book.

This book results from various suggestions from AICTE members, experts, and authors who shared their opinions and thoughts on further developing engineering education in our country. Acknowledgements are due to the contributors and different workers in this field whose published books, review articles, papers, photographs, footnotes, references, and other valuable information enriched us when writing the book.

Prof. G. C. Mohan Kumar

PREFACE

The book "Theory of Machines & Mechanisms" for III year Diploma students, as per the AICTE model curriculum, is an outcome of the teaching experience over three decades to diploma and graduate Engineering students. The initiation of writing this book is, to a great extent, exposing the mechanisms around us in day-to-day life. Even though there has been a lot of development in mechanical engineering in the last five decades, from coal-fired steam engines to diesel and electric locomotives, the principle of working and mechanisms remain unchanged. A mechanical engineer shall have a knowledge of machines and mechanisms.

The topics covered in this textbook cover under five units. Each unit is a well-defined portion of the theory of machines. The topics recommended by the AICTE model curriculum include outcome-based education and National Education Policy 2020 in a very systematic and orderly manner throughout the book. Efforts are made to explain the fundamental concepts of the mechanisms and numerical examples to understand the subject in the simplest possible way.

The book is prepared with the help of various standard textbooks available and laboratory manuals. Each chapter is concluded with a unit summary and a number of questions like multiple choice questions, as well as short and long numericals problems. The book covers all types of numerical problems that help students understand the basic retirements to solve a problem. The solution of simple numericals creates interest in analysing the role of mechanisms in conceptualising working mechanical devices.

The first unit covers concepts, definitions, and applications of cams and followers. The mechanisms for translating the rotary motion to oscillatory motion are discussed. The second unit discusses various mechanical power transmission devices, like belt drives and chain and rope drives. Additional requirements for efficient power transmission through belts are discussed. The third unit is on the smoothening of fluctuation of motion through flywheels and speed control through the governors. The fourth unit is on friction and devices like brakes, dynamometers, clutches and bearings. The last unit discusses balancing disturbances due to the imbalance of rotating masses.

Also, sources of vibrations and simple methods to solve problems with free vibrations are discussed.

I am sure that the book will help the students understand the course theory of machines in a simple way. Once students are familiar with the concept, it can be extended to solve problems of different levels. I hope this book will inspire the students to learn the course; during the course, any suggestions from readers shall be taken care to improve in the coming additions. I am happy to present the book to the student readers and the teachers.

Prof. G. C. Mohan Kumar

OUTCOME-BASED EDUCATION

For the implementation of an outcome-based education, the first requirement is to develop an outcome-based curriculum and incorporate an outcome-based assessment in the education system. By going through outcome-based assessments, evaluators will be able to evaluate whether the students have achieved the outlined standard, specific and measurable outcomes. With the proper incorporation of outcome-based education, there will be a definite commitment to achieve a minimum standard for all learners without giving up at any level. At the end of the programme running with the aid of outcome-based education, a student will be able to arrive at the following outcomes:

PO1. Basic and discipline-specific knowledge: Apply knowledge of basic mathematics, science and engineering fundamentals and an engineering specialization to solve engineering problems.

PO2. Problem analysis: Identify and analyse well-defined engineering problems using codified standard methods.

PO3. Design/development of solutions: Design solutions for well-defined technical problems and assist with designing systems components or processes to meet specified needs.

PO4. Engineering Tools, Experimentation and Testing: Apply modern engineering tools and appropriate techniques to conduct standard tests and measurements.

PO5. Engineering practices for society, sustainability and environment: Apply appropriate technology in the context of society, sustainability, environment and ethical practices.

PO6. Project Management: Use engineering management principles individually, as a team member or as a leader to manage projects and effectively communicate about well-defined engineering activities.

PO7. Life-long learning: Ability to analyse individual needs and engage in updating in the context of technological changes.

COURSE OUTCOMES

After completion of the course, the students will be able to:

CO-1: Know different machine elements and mechanisms.

CO-2: Understand kinematics and dynamics of different machines and mechanisms.

CO-3: Select suitable drives and mechanisms for a particular application.

CO-4: Appreciate the concept of balancing and vibration.

CO-5: Understand different types of cams and their motions and also draw cam profiles for various motions.

Mapping of Course Outcomes with Programme Outcomes is to be done according to the matrix given below:

Course Outcomes	Expected Mapping with Programme Outcomes (1- Weak Correlation; 2- Medium correlation; 3- Strong Correlation)						
	PO-1	PO-2	PO-3	PO-4	PO-5	PO-6	PO-7
CO-1	3	3	2	3	1	1	3
CO-2	3	3	2	3	1	1	3
CO-3	3	3	2	3	1	1	3
CO-4	3	3	2	3	1	1	3
CO-5	3	3	2	3	1	1	3

GUIDELINES FOR TEACHERS

To implement Outcome Based Education (OBE), the student's knowledge level and skill set should be enhanced. Teachers should take major responsibility for the proper implementation of OBE. Some of the responsibilities (not limited to) for the teachers in the OBE system may be as follows:

- Within reasonable constraints, they should manoeuvre time to the best advantage of all students.
- They should assess the students only upon certain defined criteria without considering any other potential ineligibility to discriminate against them.
- They should try to grow the students' learning abilities to a certain level before they leave the institute.
- They should try to ensure that all the students are equipped with quality knowledge and competence after they finish their education.
- They should always encourage the students to develop their ultimate performance capabilities.
- They should facilitate and encourage group work and teamwork to consolidate newer approaches.
- They should follow Bloom's taxonomy in every part of the assessment.

Level	Teacher should Check	Student should be able to	Possible Mode of Assessment
Create	Students ability to create	Design or Create	Mini project
Evaluate	Students ability to justify	Argue or Defend	Assignment
Analyse	Students ability to distinguish	Differentiate or Distinguish	Project/Lab Methodology
Apply	Students ability to use information	Operate or Demonstrate	Technical Presentation/ Demonstration
Understand	Students ability to explain the ideas	Explain or Classify	Presentation/Seminar
Remember	Students ability to recall (or remember)	Define or Recall	Quiz

GUIDELINES FOR STUDENTS

Students should take equal responsibility for implementing the OBE. Some of the responsibilities (not limited to) for the students in the OBE system are as follows:

- Students should be well aware of each UO before the start of a unit in each and every course.
- Students should be well aware of each CO before the start of the course.
- Students should be well aware of each PO before the start of the programme.
- Students should think critically and reasonably with proper reflection and action.
- Learning of the students should be connected and integrated with practical and real-life consequences.
- Students should be well aware of their competency at every level of OBE.

ABBREVIATIONS AND SYMBOLS

a	acceleration m/s^2	μ	coefficient of friction
b	breadth m	δ	change, difference
c	coefficient %	β	angle in rad
C	constant	σ	stress Pa
d	diameter, distance m	θ	angular rotation, lap angle rad
E	energy J, Young's modulus Pa	ω	speed rad/s
e	eccentricity m	SHM	simple harmonic equation
F	force N	VR	velocity ratio
G	gravity m/s^2	IC	internal combustion
H	height m		
I	mass moment of inertia kgm^2		
K	radius of gyration m		
k	stiffness N/m		
L	displacement m, length m		
M	mass kg, module mm		
N	speed rpm		
P	force N		
P	power watt		
q	ratio of masses		
R	effective radius m		
r	radius m		
R	reaction N		
s	slip %		
S	spring balance reading		
T	Number of teeth, Turning moment Nm Tension N Torque Nm		
t	thickness m time sec		
v	velocity m/s		
W	weight N		
x	distance m		
X	distance m		
α	angle of lap rad		
π	Pi		

LIST OF FIGURES

LIST OF TABLES

CONTENTS

1 CAMS AND FOLLOWERS

UNIT SPECIFICS

This unit presents the concept and application of Cams and Followers. The unit comprehensively overviews the types of followers and cam profiles. The displacement of followers and their velocity and acceleration for the given cam profiles will be studied. In particular, to uniform velocity, SHM, uniform acceleration and Retardation. The student is able to draw the profile of a radial cam with knife-edge and roller follower with and without offset with reciprocating motion using a graphical method.

RATIONALE

The repetitive motion of parts that are used in machinery uses cams to translate from rotary to reciprocating. The knowledge of cams and followers is essential for mechanical engineers in designing and manufacturing cams.

PRE-REQUISITE

Nil

UNIT OUTCOMES

The list of outcomes of this unit is as follows:
U1-O1: Understand the concept and application of Cams and Followers.
U1-O2: To draw the Displacement, velocity and acceleration diagrams for a follower.
U1-O3: To draw the cam profile according to the displacement of the follower.

Unit Outcomes	**Expected Mapping with the Course Outcomes** (1- Weak Correlation; 2- Medium Correlation; 3- Strong Correlation)				
	CO-1	CO-2	CO-3	CO-4	CO-5
U1-O1	3	3	3	1	3
U1-O2	3	3	3	1	3
U1-O3	3	3	3	1	3

1.1 Concept, Definition, and Applications

The cam and follower mechanisms are extensively used in mechanical equipment to produce translational and oscillatory motion from a rotating shaft. This mechanism is used whenever continuous or periodic motion is required, like driving the inlet and outlet valves in internal combustion engines, fuel pumps, machine tools, etc. The cam and follower mechanism has a line contact, constituting a higher pair. A cam is mounted on a shaft which drives the follower to give a specified displacement and time. The cam's shape depends on the follower's displacement with respect to time in a given rotation. The shaft carrying a cam or many cams is called a camshaft. As it rotates, it lifts the follower to the predefined distance. This is the simplest and one of the most important mechanisms used to transform the motion from rotary into linear and is found in modern machinery today.

Applications of cam and followers are also found in spinning and weaving textile machinery, automatic attachment for machinery, paper cutting machines, cutting tool feed mechanisms of automatic lathes, etc. The combination of cam and follower has the following elements.

- A driver shaft carrying a cam on it.
- A follower held securely against a spring force to always keep contact with the cam.
- A frame to accommodate the cam and follower guideways.

1.2 Classification of Cams and Followers

The cams are classified as a) Radial cams, b) Cylindrical cams, and c) Special cams.

a) Radial cams are very popular among all types of cams. This type of cam is used when the follower needs to reciprocate radially, as shown in Fig. 1.1(a). These cams are simple in geometry, easy to manufacture, and compact.

b) In cylindrical cams, a groove is cut on the surface, which guides the follower to move either reciprocating or oscillating, as shown in Fig. 1.1(b)

c) The special cams largely depend on the type of follower path required. Such special cams include wedge and follower, spiral, spherical, and conjugate. The Globoidal cams are used where the follower requires a large oscillation.

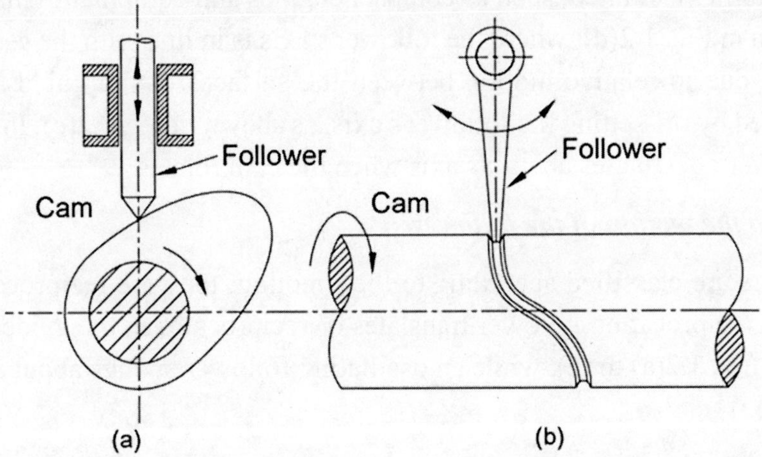

Fig.1.1 Types of cams: (a) Radial cam and (b) Cylindrical cam.

The followers used along with cams are classified according to the surface of the contact surface, the motion and the path of motion of the follower. The different types of followers used are discussed here.

According to the contact surface:

- **Knife edge follower:** According to the classification based on the contact surface, a follower is identified as a knife edge follower whose contact end has a sharp knife edge and is shown in Fig. 1.2(a). Knife edge followers are limited due to excessive wear due to the sliding surfaces of the cam and follower.

- **Roller follower:** This type of follower has a roller at its end that rolls over the cam's surface, as shown in Fig.1.2(b). The rolling motion of the cam and follower offers reduced wear of both cam and follower. The roller follower is widely used in oil engines to operate inlet and exhaust valves. This type of roller follower is very commonly used in energy-efficient machinery.

- **Spherical-faced follower:** The high surface stresses produced in the flat surface of the follower can be minimised by using a spherical shape. The contacting end of the follower is spherical shaped as shown in Fig. 1.2(c).
- **Flat-faced follower:** The contact end of the follower is a circular flank with a flat face and is called a flat-faced follower. The flat-faced followers can be used where the space

availability is limited, such as compact engines and equipment. This type of follower is shown in Fig. 1.2(d), where the follower's axis is in line with the centre of the cam. The thrust due to relative motion between the surfaces of the guide and follower can be reduced by off-setting the follower's axis, as shown in Fig.1.2(e). In this mechanism, the follower also rotates about its axis when the cam rotates.

According to the motion of the follower:

The followers are classified according to their motion; they are reciprocating and oscillating followers. A reciprocating follower translates or reciprocates in the guides as the cam rotates, as shown in Fig. 1.2(a) to (e), while an oscillating follower swings about a point and is shown in Fig. 1.2(f).

According to the location of the follower:

According to the follower's location concerning the axis cam, the followers are of two types: Radial and Offset followers.

- **Radial follower:** The radial line of the cam and the line of motion of the follower are the same, known as a radial follower. Fig. 1.2(a-d) shows the radial followers.

- **Off-set follower:** When the motion of the follower is away from the axis of the cam centre, such a follower is called an off-set follower. This type of follower is shown in Fig. 1.2(e).

1.3 Definition

A simple radial cam with a reciprocating roller follower is shown in Fig.1.3. The essential terms generally used in developing the cam profile are discussed.

Base circle: The smallest circle can be drawn to the cam profile.

Pitch point: It is a point on the pitch curve having the maximum pressure angle.

Trace point: A reference point on the follower used to generate the pitch curve.

Pitch curve: A curve generated by the tracepoint as the follower moves relative to the cam.

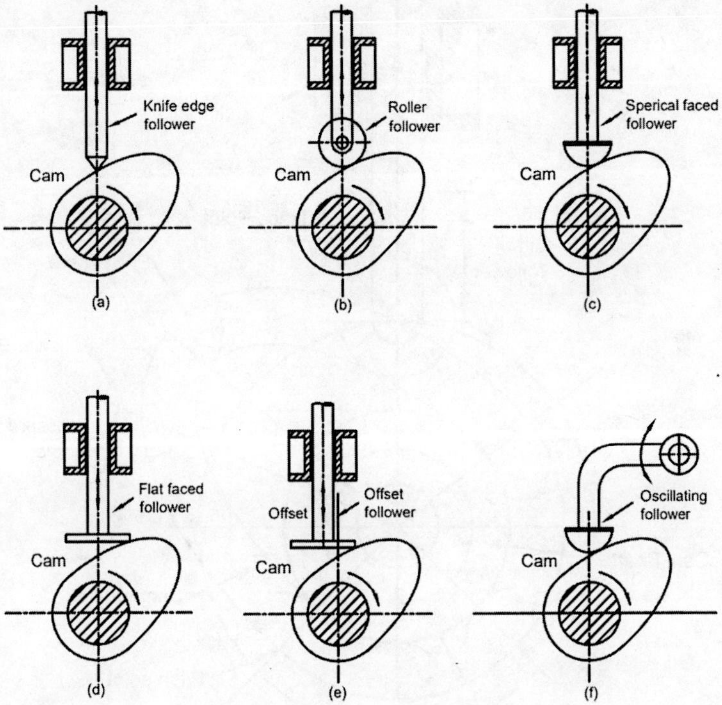

Fig.1.2 Type of followers.

Pressure angle: The angle between the direction of the follower motion and a normal to the pitch curve.

Pitch circle: A circle drawn from the centre of the cam through the pitch points.

Prime circle: It is the smallest circle drawn from the centre of the cam and tangent to the pitch curve.

Lift or stroke: The maximum travel of the follower from its lowest to the highest position in one rotation.

Angle of outstroke: The angle or rotation of the cam required to complete the outward stroke of the follower.

Angle of return stroke: The angle or rotation of the cam required to complete the return stroke of the follower.

Angle of dwell: The angle or rotation of the cam required to keep the follower displacement constant or unmoved.

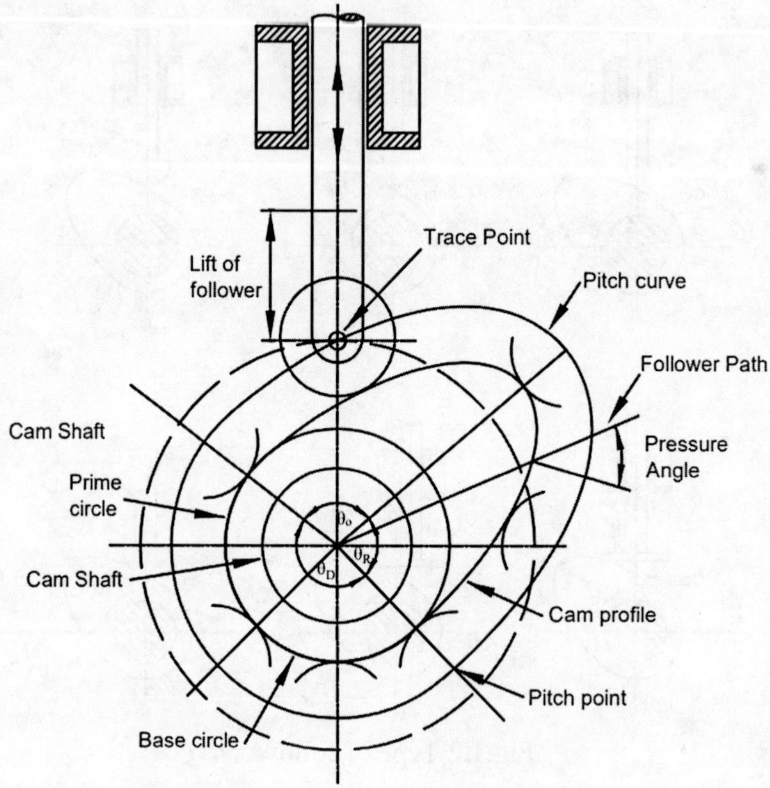

Fig.1.3 Terms used in cam and follower.

1.4 Motion of a Follower

During function, the rotating cam drives the follower to reciprocate. The motion of the follower has a specific type of motion during the out stroke and return stroke. They may have any one type of motion, as listed below.

1. Uniform velocity,
2. Uniform acceleration and retardation,
3. Simple harmonic motion and others.

As the cam rotates, a follower's displacement, Velocity and Acceleration vary continuously. The predefined displacement of the follower gives the required velocity and acceleration. A displacement diagram for the follower helps to generate the cam profile. A knowledge of displacement, velocity and acceleration is essential from the dynamics point of view for the smooth functioning of the cam and follower mechanism.

Displacement of the follower and angular position of the cam:

There are four segments in the displacement of the follower. They are a) outstroke, b) dwell, c) return stroke and d) dwell.

As the cam begins to rotate, the follower gradually rises from zero displacement to maximum displacement 'L' during '**outstroke**' in an angular rotation θ_1 of the cam, as shown in Fig. 1.4. After the rise, there will be '**dwell**' of the follower, where the follower has no movement, and this occurs for an angular rotation θ_2 of the cam. During the '**return stroke**' of the follower, it reaches zero displacement position with the angular rotation θ_3 of the cam. The remaining angular rotation θ_4 of the cam follows '**dwell**' as shown in figure. Sometimes, the dwell periods like θ_2 or θ_4 may also be equal to zero. All rise, return and dwell periods depend on the motion of the follower required for a particular application. In one full cam rotation, the follower has been lifted to the height of L during the outstroke and returned to its original position with the return stroke. The following is the position of follower in one revolution of cam,

1. $L = 0$, at point A when $\theta = 0$
2. $L =$ maximum at point B' during out stroke $\theta = \theta_1$
3. $L =$ maximum and constant till point C' during dwell $\theta = \theta_2$
4. $L = 0$ at point D, during return stroke $\theta = \theta_3$
5. $L = 0$ and remains constant during dwell till E, $\theta = \theta_4$

Let ω be the speed or angular velocity of the cam. An angular rotation of the cam during the outstroke of the follower θ_1 and for the return stroke is θ_3. The angular rotation of the cam for the dwell is θ_2 and θ_4. The time required for each activity is given by

$t = \theta_1/\omega$ for outstroke

$t = \theta_2/\omega$ for dwell - 1

$t = \theta_3/\omega$ for return stroke and

$t = \theta_4/\omega$ for dwell - 2

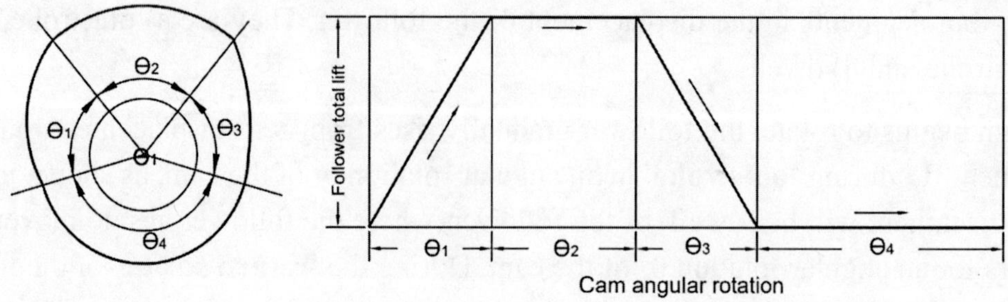

Fig.1.4 Follower lift with cam rotation.

1.4 a) Follower with uniform velocity motion

In this case, the displacement of the follower is linear with the rotation of the cam, and therefore, it has a uniform velocity during its outstroke and return stroke. In other words, the slope of follower displacement is uniform. The acceleration or retardation of the follower is infinite at the beginning and at the end of each stroke. For a knife-edged follower, the variation of displacement, velocity and acceleration for the case of uniform velocity is shown in Fig. 1.5(a). For the remaining dwell periods, the follower remains at rest.

The follower moves with uniform velocity from initial position A to B during its outstroke with the rotation of shaft θ_1. Further, the position of the follower is at rest during dwell B'-C' with the rotation of shaft θ_2. In the return stroke C-D, the follower with uniform velocity descends as the cam rotates through an angle θ_3. For the remaining cam rotation θ_4, the follower remains at rest as dwell period D-E, as shown in Figure. The acceleration and retardation at the beginning and end of each rise and return of the follower suffer from infinite values. To avoid such motion, the motion of the follower is modified by rounding off the sharp corners at the displacement curve, as shown in Fig. 1.5(b). Due to this, the velocity of the follower varies gradually at the beginning and end of each stroke. The rounded corners are usually parabolic curves in the displacement diagram. The parabolic motion of the follower at the beginning and end of each outstroke and return results in a very low and finite follower acceleration. The displacement, velocity and acceleration of a knife edge follower for uniform velocity without and with curves are shown in Fig. 1.5.

Let L be the total displacement of the follower, and it has been achieved during the outstroke with time $t = \theta_1/\omega$.

Therefore, the velocity of the follower $V_O = L/(\theta_1/\omega)$

and similarly, for the return stroke, the velocity $V_R = L/(\theta_2/\omega)$.

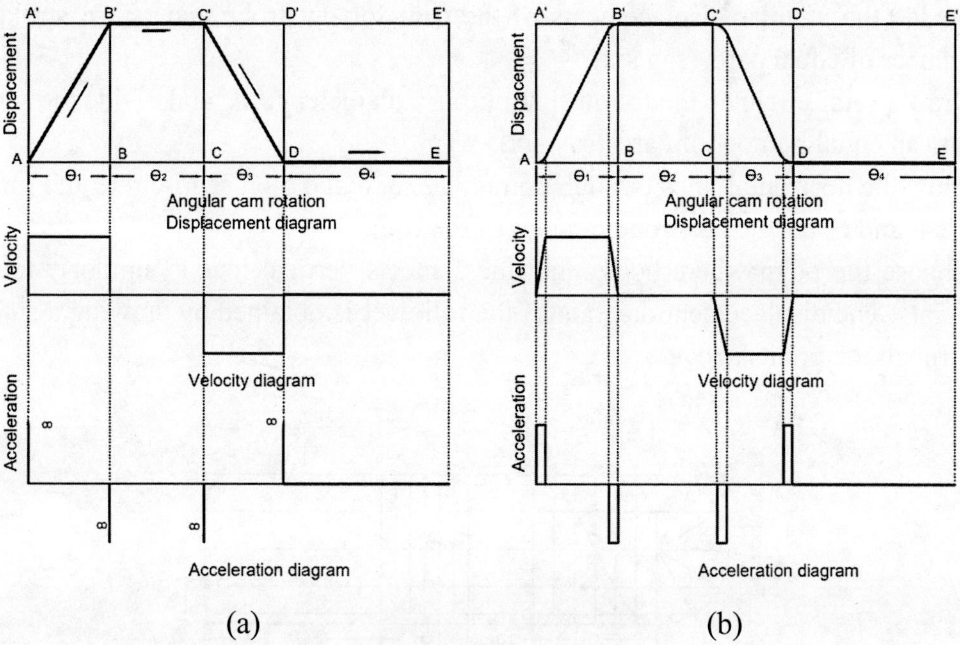

Fig.1.5 Displacement, velocity and acceleration diagrams for knife edge follower (a) simple cam and (b) modified cam profile.

1.4 b) Follower with Simple Harmonic Motion (SHM)

The follower rises or descends with a simple harmonic motion. The lift or height of the follower is equal to the projected distance over the diagonal of a semi-circle. Fig.1.6 shows the displacement diagram when the follower moves with simple harmonic motion. Various points on the semicircle are projected over the diagonal. This gives the displacement of the follower from zero to its maximum, as shown along A A' in the figure. The follower's velocity follows a sine curve, while its acceleration is a cosine curve.

As shown in Fig.1.6, the velocity of the follower is zero at the beginning and increases gradually to a maximum at mid-stroke. Again, at the end of this stroke, the velocity of the follower is zero. The acceleration of the follower is maximum at the beginning. It diminishes to zero at mid-stroke, and at the end of the stroke, the acceleration is zero, as shown in the figure. Similarly, during the follower's return stroke, the displacement, velocity, and acceleration when the follower returns with simple harmonic motion are shown in the same figure. The displacement diagram for the follower for SHM is drawn as follows:

1. Set the angular position of the cam according to the period of rise, dwell and return. The diagonal of a semicircle is equal to the maximum displacement.
2. Divide the angular displacements of the cam for outstroke and return strokes into a number of equal parts (say six).
3. Draw a semi-circle on the follower stroke as a diameter - AA' and divide the semi-circle into an equal number of parts (say six).
4. Draw the horizontal lines through points 1, 2, 3, 4 and 5 on semi-circle and lines 1', 2', 3', 4' and 5' for the outstroke rotation of the cam.
5. Choose the point where horizontal line 1 meets vertical line 1' similarly to all other points. The displacement diagram of the follower is obtained by drawing a smooth line through the points above.

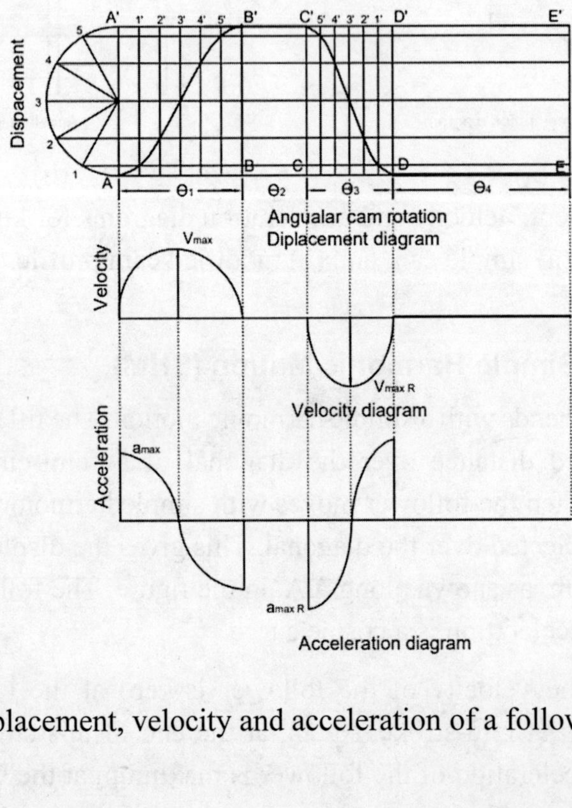

Fig.1.6 Displacement, velocity and acceleration of a follower with SHM.

Velocity and acceleration of a follower

Let L = displacement of the follower in mm,

θ = Angular rotation of the cam during outstroke in radians and

ω = Angular velocity of the cam in rad/s.

Time taken for outstroke by the follower in seconds

$$T = \theta/\omega$$

Consider a contact point of a follower with a cam moving at a uniform speed ω radians per sec (around the circumference of a circle with stroke L as diameter), as shown in the figure. When a distance (πL/2) is covered in time T, then the velocity is given by

$$v = \frac{\pi L}{2T} = \frac{\pi \omega L}{2\theta}$$

The acceleration for the above is given by (v^2/distance)

$$a = \left(\frac{\pi \omega L}{2\theta}\right)^2 \left(\frac{2}{L}\right) = \left(\frac{\pi \omega}{\theta}\right)^2 \left(\frac{L}{2}\right)$$

The above expressions for the velocity and accelerations can be used for a follower for the return stroke.

1.4 c) Follower with Uniform Acceleration and Retardation

This type of cam and follower mechanism has a uniform acceleration during the first half of the stroke. For the remaining part of the return stroke, there is a uniform retardation of the follower. The displacement diagram of the follower consists of a parabolic curve and can be generated, as explained in the following section. Fig.1.7 shows the displacement, velocity and acceleration of the follower when it moves outstroke and return stroke with uniform acceleration and retardation, respectively.

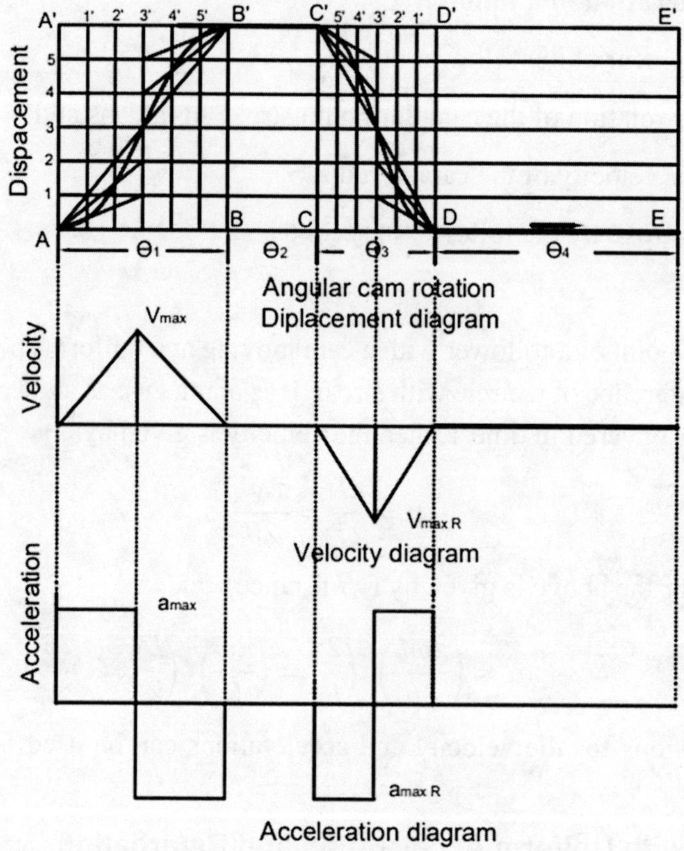

Fig.1.7 Displacement, velocity and acceleration of a follower with constant acceleration and deacceleration.

The velocity of the follower is zero at the beginning and end of the stroke. However, the maximum velocity of the follower is observed in the middle of the stroke. The velocity is negative or reversed during the return stroke, as shown in the figure. The acceleration of the follower begins with a jump and is constant for the first half, and a sudden change in acceleration in the middle of the stroke is observed. Further, the remaining part of the stroke is negative or reversed and is again constant till the end of the out stroke. A similar behaviour of the follower during return stroke is observed. In the beginning, it was negative and constant. Further, as it progresses, at the middle of the return stroke, a change to deceleration occurs, and again, it is constant, as shown in the figure.

Velocity and acceleration of a follower

Let L = displacement of the follower in mm,

θ = Angular rotation of the cam during outstroke in radians and

ω = Angular velocity of the cam in rad/s.

Time taken for outstroke by the follower in seconds

$$T = \theta/\omega$$

The velocity is given by (L/T), and the maximum velocity of the follower occurs at ($\theta_1/2$), as shown in the figure.

$$v = \frac{2\omega L}{\theta}$$

The maximum velocity of the follower is reached at the midpoint of the outstroke. During the first half of the outstroke, there is a uniform acceleration, as shown in the acceleration diagram. The next half of the outstroke is a uniform retardation. Then, the maximum acceleration of the follower during the outstroke,

$$a = \frac{v}{T/2} = 2\frac{2\omega L}{\theta\left(\frac{\theta}{\omega}\right)}$$

$$or \quad a = \left(\frac{2\omega}{\theta}\right)^2 L$$

The above expressions for the velocity and accelerations of the follower can be used in motion during a return stroke.

1.5 Construction of Cam Profile

The cam profiles largely depend on the type of motion, radial or offset and the type of follower used. Considering all these, a basic displacement diagram for the given motion of the follower is sketched. The displacement diagram provides an accurate follower displacement from its base circle with the angular position of the cam. Its radial position at each angular position is translated on the cam to locate the points on the proposed cam profile. The working surface of the cam is nothing but the locus of these points, and the cam profile is created by using a smooth contour. While constructing the cam profile, the basic principle of kinematic inversion is utilised. i.e. the cam is assumed to be stationary or fixed while the follower is imagined to revolve in the opposite direction to the cam rotation. The step-by-step procedure for creating a cam profile is given in the following section. The cam profiles are developed for the following cases,

- Type of follower: knife edge, roller and flat type.
- Location of Follower: radial and offset.
- Motion of Follower: constant velocity, acceleration/deacceleration and SHM.

Example 1: Draw a cam profile for the following motion of a knife-edged follower.

Outstroke of follower during 90° of cam rotation. Dwell for the next 60° of cam rotation. Return stroke during the next 90° of cam rotation. Dwell for the remaining 120° of cam rotation. The stroke of the follower is 60 mm, and the minimum radius of the cam is 40 mm. The follower has a uniform velocity motion during the outstroke and return strokes. The axis of the follower passes through the axis of the camshaft.

Solution:

Construction: Displacement Diagram

The displacement diagram for the above cam is shown in Fig.1.8(a). The following steps are used to create the displacement diagram.

- Draw a rectangle representing height as a displacement 60 mm and length to one rotation of cam = 360° to some suitable scale. AA'=60 mm and AE= 100 mm.
- Divide the rectangle into four blocks representing outstroke, dwell, return stroke and dwell in proportion.
- Divide the angular displacement during outstroke and return stroke into an equal number of six parts and draw vertical lines through each point as shown.

- The follower moves with **uniform velocity** during the outstroke and return stroke, and the displacement of the follower is proportionate with the angular position. Draw a straight-line AB' for outstroke and C'D for return stroke to represent the displacement diagram.

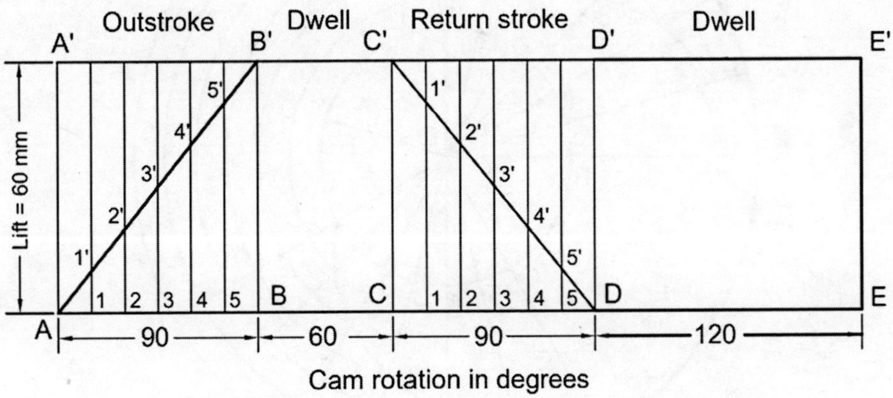

Fig.1.8(a) Displacement Diagram for Example 1.

Construction of Cam Profile: To sketch the cam profile, the following steps are used and are shown in Fig.1.8(b).

- Draw a base circle with a radius OA=40 mm with O as the centre.
- With A as a trace point, sketch the knife edge follower vertically A-A'. The axis of the follower shall pass through the axis of the camshaft.
- Divide the circle into four parts according to outstroke, dwell, return stroke and dwell from OA, as shown in the figure. A'OB' = 90°, B'OC'=60° and C'OD'=90°. Further, subdivide the outstroke and return stroke angle sector into six equal parts. Draw a radial line from centre O.
- Transfer the lengths 1-1', 2-2' and so on from the displacement diagram and mark points 1', 2', 3',4', and 5' for out stroke and 5', 4' … 1' for return stroke.
- Join the points A, 1', 2', 3',4', 5' and B' for out stroke and B', 5', 4', 3', 2', 1', and D for return stroke with a smooth curve. Dwell portions B' to C' and D to A are the arcs, as shown in the figure. This is the profile of the cam required.

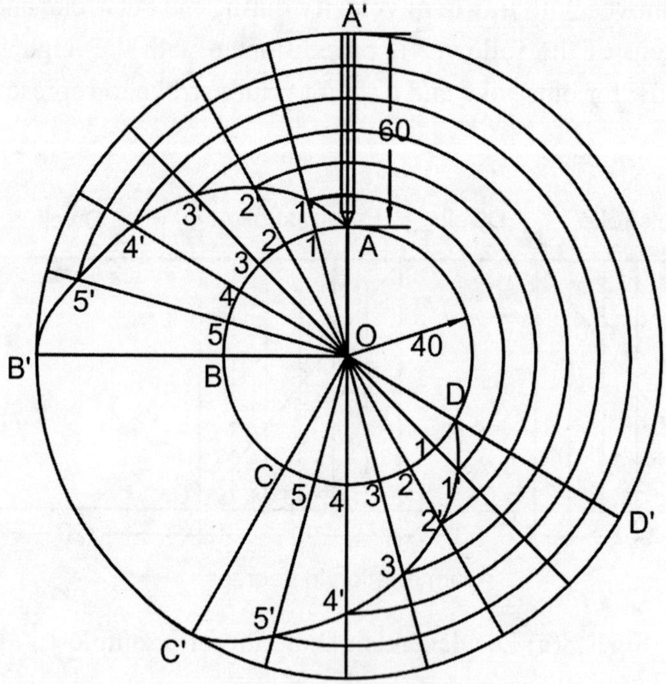

Fig.1.8(b) Cam profile for the for Example 1.

Example 2: Draw a cam profile for the following motion to the knife-edged follower with an offset of 15mm from the cam centre.

Outstroke of follower during 90° of cam rotation. Dwell for the next 60° of cam rotation. Return stroke during the next 90° of cam rotation. Dwell for the remaining 120° of cam rotation.

The stroke of the follower is 70 mm, and the minimum radius of the cam is 50 mm. The follower has a uniform velocity motion during the outstroke and return strokes. The axis of the follower passes through the axis of the camshaft.

Solution:

Construction: Displacement Diagram

The displacement diagram for the above cam is shown in Fig. 1.9(a). The following steps are used to create the displacement diagram.

- Draw a rectangle representing height as a displacement and length to one rotation of cam = 360° to some suitable scale. AA'=70 mm and AA= 100 mm.

- Divide the rectangle into four blocks representing outstroke, dwell, return stroke and dwell in proportion.
- Divide the angular displacement during outstroke and return stroke into an equal number of six parts and draw vertical lines through each point as shown.
- The follower moves with **uniform velocity** during the outstroke and return stroke, and the displacement of the follower is proportionate with the angular position. Draw a straight-line AB' for outstroke and C'D for return stroke to represent the displacement diagram.

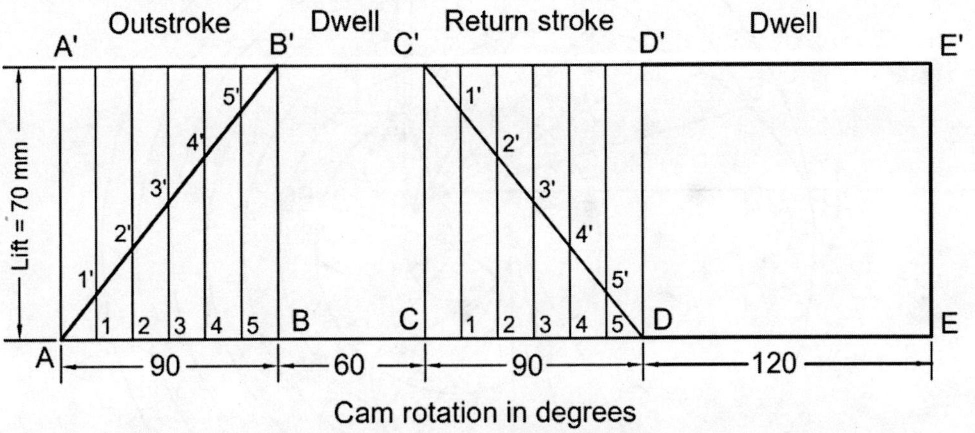

Fig.1.9(a) Displacement Diagram for Example 2.

Construction of Cam Profile: To sketch the cam profile, the following steps are used and are shown in Fig.1.9(b).

- Draw a base circle with a radius of OA=50 mm and another circle with a radius of 15 mm with O as the centre. Let the axis of the follower be offset by 15 mm and tangent to offset the circle, as shown in the figure.
- With tracepoint A, draw a line OA. Divide the circle into four parts according to outstroke, dwell, return stroke and dwell from OA, as shown in the figure. Angle AOB = 90°, angle BOC = 60° and angle COD = 90°. Further, subdivide the outstroke and return stroke angle sector into six equal parts. Let points 1, 2, 3, 4, and 5 be on the base circle for outstroke, and 5, 4, 3, 2, and 1 for return stroke. For both strokes, draw tangent lines to offset the circle from points 1 to 5 and B.

- Transfer the lengths 1-1', 2-2' and so on from the displacement diagram and mark points 1', 2', 3',4', and 5' for out stroke and 5', 4' … 1' for return stroke.
- Join the points A, 1', 2', 3',4', 5' and B' for out stroke and C', 5', 4', 3', 2', 1', and D for return stroke with a smooth curve. Dwell portions B' to C' and D to A are the arcs, as shown in the figure. This is the profile of the cam required.

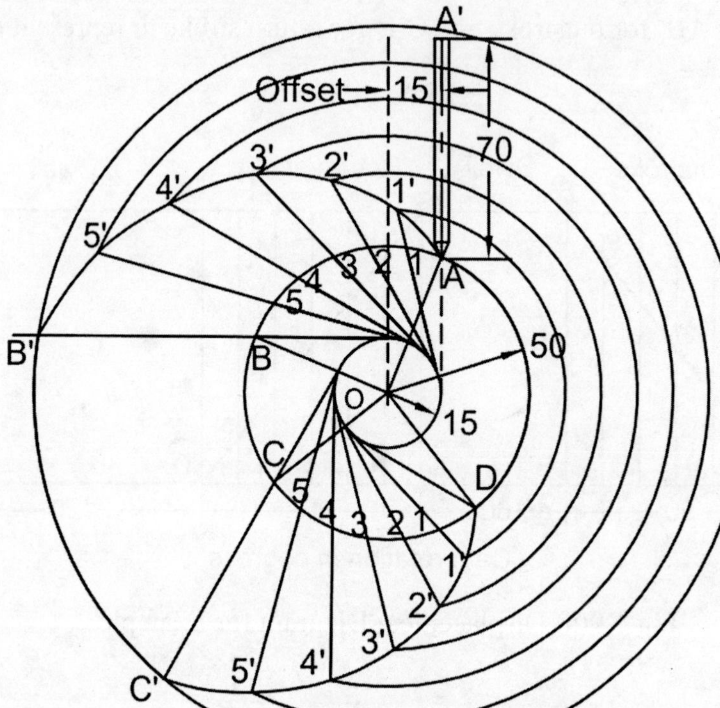

Fig.1.9(b) Cam profile for Example 2.

Example 3: Draw a cam profile for the following motion to a roller follower with a diameter of 20 mm. Outstroke during 90° of cam rotation. Dwell for the next 60° of cam rotation. Return stroke during the next 90° of cam rotation. Dwell for the remaining 120° of cam rotation.

The stroke of the follower is 50 mm, and the minimum radius of the cam is 40 mm. The follower has a uniform velocity motion during the outstroke and return strokes. The axis of the follower passes through the axis of the camshaft.

Solution:

Construction: Displacement Diagram

The displacement diagram for the above cam is shown in Fig.1.10(a). The following steps are used to create the displacement diagram.

- Draw a rectangle representing height as a displacement and length to one rotation of cam = 360° to some suitable scale. AA'=50 mm and AA= 100 mm.
- Divide the rectangle into four blocks representing outstroke, dwell, return stroke and dwell in proportion.
- Divide the angular displacement during outstroke and return stroke into an equal number of six parts and draw vertical lines through each point as shown.
- The follower moves with **uniform velocity** during the outstroke and return stroke, and the displacement of the follower is proportionate with the angular position. Draw a straight-line AB' for outstroke and C'D for return stroke to represent the displacement diagram.

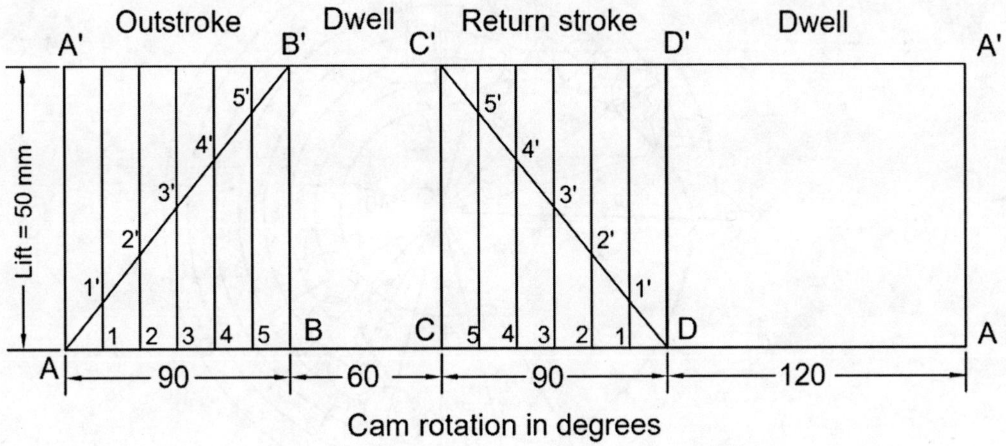

Fig.1.10(a) Displacement Diagram for Example 3.

Construction of Cam Profile: To sketch the cam profile, the following steps are used and are shown in Fig.1.10(b).

- Draw a base circle with the centre with a radius OA= 40 mm.
- Locate the centre of the roller above the centre of the cam.; base circle radius + radius of roller = 40 + 10 = 50mm.

- Divide the base circle into four parts according to outstroke, dwell, return stroke and dwell from OA, as shown in the figure. A’OB’ = 90°, B’OC’=60° and C’OD’=90°. Further, subdivide the outstroke and return stroke angle sector into six equal parts. Draw a radial line from centre O.
- Transfer the lengths 1-1’, 2-2’ and so on from the displacement diagram and add the radius of the roller. Mark points 1’, 2’, 3’,4’, and 5’ for out stroke and 5’, 4’ … 1’ for return stroke.
- Draw the circle representing the roller with centre 1’, 2’,…5’ for out stroke and similarly for return stroke.
- Draw a smooth curve tangential to all the circles with centres A, 1’, 2’, 3’,4’, 5’ and B’ for out stroke and B’, 5’, 4’, 3’, 2’, 1’, and D for return stroke with a smooth curve. Dwell portions B’ to C’ and D to A are the arcs, as shown in the figure. This is the profile of the cam required.

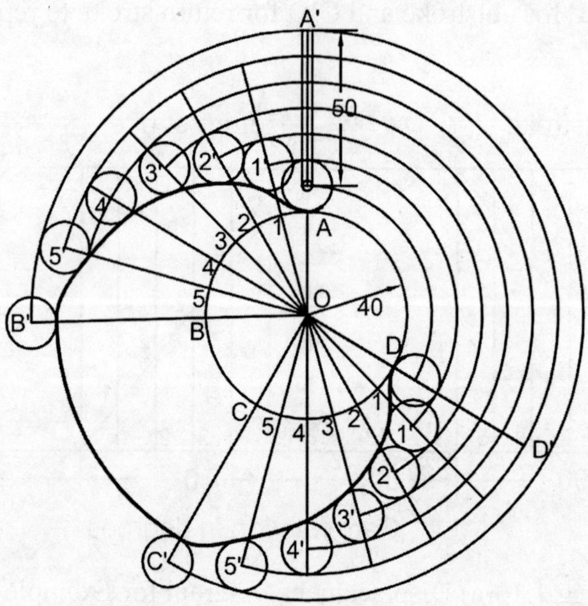

Fig.1.10(b) Cam profile for Example 3.

Example 4: Draw a cam profile with a knife-edged follower. The outstroke during 120° of cam rotation, dwell for the next 60° of cam rotation and return stroke during the next 90° of cam rotation is followed. Dwell for the remaining 90° of cam rotation.
The stroke of the follower is 20 mm, and the minimum radius of the cam is 50 mm. The follower has a Simple harmonic motion during the outstroke and return strokes. The axis of the follower passes through the axis of the camshaft.

Solution:

Construction: Displacement Diagram

The displacement diagram for the above cam is shown in Fig.1.11(a). The following steps are used to create the displacement diagram.

- Draw a rectangle representing height as a displacement and length to one rotation of cam = 360° to some suitable scale. AA' = 20 mm and AA = 100 mm.
- Divide the rectangle into four blocks representing outstroke, dwell, return stroke and dwell in proportion.
- Divide the angular displacement or rotation during the outstroke and return stroke into an equal number of six parts and draw vertical lines through each point as shown.
- Draw a semicircle on line AA' and divide the semicircle into six equal parts. Mark points P1 to P5, as shown in the figure below. The follower displacement for simple harmonic motion is obtained by projecting points P1, P2,… P5 horizontally to a particular angular position to get points 1', 2', ….5' for outstroke and return stroke. Draw a smooth curve to obtain displacement curves A to B' and C' to D, as shown in the figure.

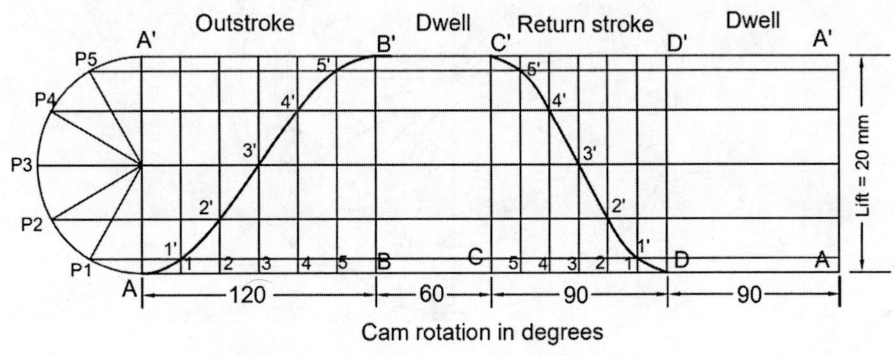

Fig.1.11(a) Displacement Diagram for Example 4.

Construction of Cam Profile: To sketch the cam profile, the following steps are used and are shown in Fig.1.11(b).

- Draw a base circle with a radius of OA=50 mm and a follower vertically above, as shown in the figure. Over the follower, draw a semicircle with a diameter of 20mm.
- Divide the semicircle into 6 equal parts and draw the line projected on diameter as described in the follower displacement diagram. Draw circles through these points from the cam centre.
- Divide the circle into four parts according to outstroke, dwell, return stroke and dwell from OA, as shown in the figure. Angle A'OB' = 120°, angle B'OC'=60° and angle C'OD'=90°. Further, subdivide the outstroke and return stroke angle sector into six equal parts. Let points 1, 2, 3, 4, and 5 be on the base circle for outstroke, and 5, 4, 3, 2, and 1 for return stroke. Draw the radial lines of these points from the centre.
- Transfer the lengths, 1-1', 2-2' and so on from the displacement diagram and mark points 1', 2', 3',4', and 5' for out stroke and 5', 4' … 1' for return stroke over the base circle. As shown in the figure below, these points can also be located with radial lines corresponding to the angular position and displacement circles.
- Join the points A, 1', 2', 3',4', 5' and B' for out stroke and C', 5', 4', 3', 2', 1', and D for return stroke with a smooth curve. Dwell portions B' to C' and D to A are the arcs, as shown in the figure. This is the profile of the cam required.

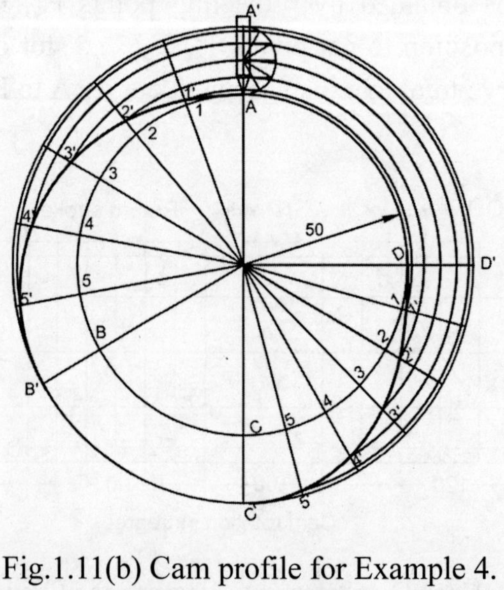

Fig.1.11(b) Cam profile for Example 4.

Example 5: Draw a displacement diagram and cam profile with a knife-edged follower. The outstroke during 120° of cam rotation, dwell for the next 60° of cam rotation and return stroke during the next 90° of cam rotation is followed. Dwell for the remaining 90° of cam rotation. The stroke of the follower is 30 mm, and the minimum radius of the cam is 50 mm. The follower has SHM during the outstroke and constant acceleration and deceleration during the return stroke. The axis of the follower passes through the axis of the cam.

Solution:

Construction: Displacement Diagram

The displacement diagram for the above cam is shown in Fig. 1.12(a). The following steps are used to create the displacement diagram.

- Draw a rectangle representing height as a displacement and length to one rotation of cam = 360° to some suitable scale. AA'=30 mm and AA= 100 mm. Divide the rectangle into four blocks representing outstroke, dwell, return stroke and dwell in proportion.
- Divide the angular displacement or rotation during the outstroke and return stroke into an equal number of six parts and draw vertical lines through each point as shown.
- Draw a semicircle on line AA' and divide the semicircle into six equal parts. Mark points P1 to P5, as shown in the figure below. Draw horizontal lines through these points to get points 1', 2', ...5' for outstroke.
- For the return stroke, draw lines from point C' to Q1, Q2 and Q3 and D to Q4 and Q5. Mark points 5' to 1' as shown in the figure. Draw a smooth curve to obtain displacement curves A to B' and C' to D, as shown in the figure.

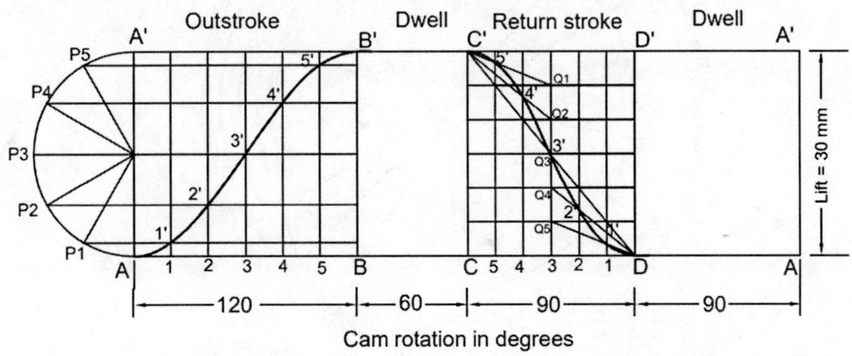

Fig.1.12(a) Displacement Diagram for Example 5.

Construction of Cam Profile: To sketch the cam profile, the following steps are used and are shown in Fig.1.12(b).

- Draw a base circle with a radius of OA=50 mm and a follower vertically above, as shown in the figure. Over the follower, draw a semicircle with a diameter of 30 mm.
- Divide the semicircle into 6 equal parts and draw the line projected on diameter as described in the follower displacement diagram. Draw circles through these points from the cam centre.
- Divide the cam circle into four parts according to outstroke, dwell, return stroke and dwell from OA, as shown in the figure. Angle A'OB' = 120°, angle B'OC'=60° and angle C'OD'=90°. Further, subdivide the outstroke and return stroke sectors into six equal parts. Let points 1, 2, 3, 4, and 5 be on the base circle for outstroke, and 5, 4, 3, 2, and 1 for return stroke. Draw the radial lines of these points from the centre.
- Transfer the lengths, 1-1', 2-2' and so on from the displacement diagram for SHM outstroke and constant acceleration and deacceleration motion for return stroke over the base circle.
- Join the points A, 1', 2', 3',4', 5' and B' for out stroke and C', 5', 4', 3', 2', 1', and D for return stroke with a smooth curve. Dwell portions B' to C' and D to A are the arcs, as shown in the figure. This is the profile of the cam required.

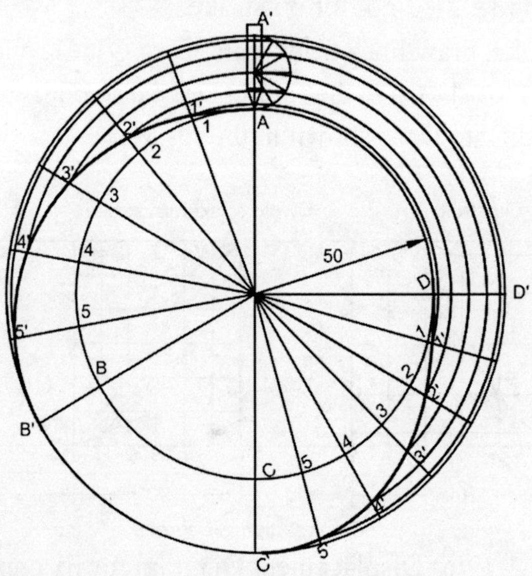

Fig.1.12(b) Cam profile for Example 5.

Example 6: Draw a displacement diagram and cam profile with a knife-edged follower. The follower's axis is offset by 10 mm from the centre of the cam. The outstroke during 120° of cam rotation, dwell for the next 60° of cam rotation and return stroke during the next 90° of cam rotation is followed. Dwell for the remaining 90° of cam rotation. The stroke of the follower is 30 mm, and the minimum radius of the cam is 50 mm. The follower has SHM during the outstroke and constant acceleration and deceleration during the return stroke.
Solution:

Construction: Displacement Diagram

The displacement diagram for the above cam is shown in Fig. 1.13(a). The following steps are used to create the displacement diagram.

- Draw a rectangle representing height as a displacement and length to one rotation of cam = 360° to some suitable scale. AA'=30 mm and AA= 100 mm. Divide the rectangle into four blocks representing outstroke, dwell, return stroke and dwell in proportion.
- Divide the angular displacement or rotation during the outstroke and return stroke into an equal number of six parts and draw vertical lines through each point as shown.
- Draw a semicircle on line AA' and divide the semicircle into six equal parts. Mark points P1 to P5, as shown in the figure below. Draw horizontal lines through these points to get points 1', 2', ...5' for outstroke.
- For the return stroke, draw lines from point C' to Q1, Q2 and Q3 and D to Q4 and Q5. Mark points 5' to 1' as shown in the figure. Draw a smooth curve to obtain displacement curves A to B' and C' to D, as shown in the figure.

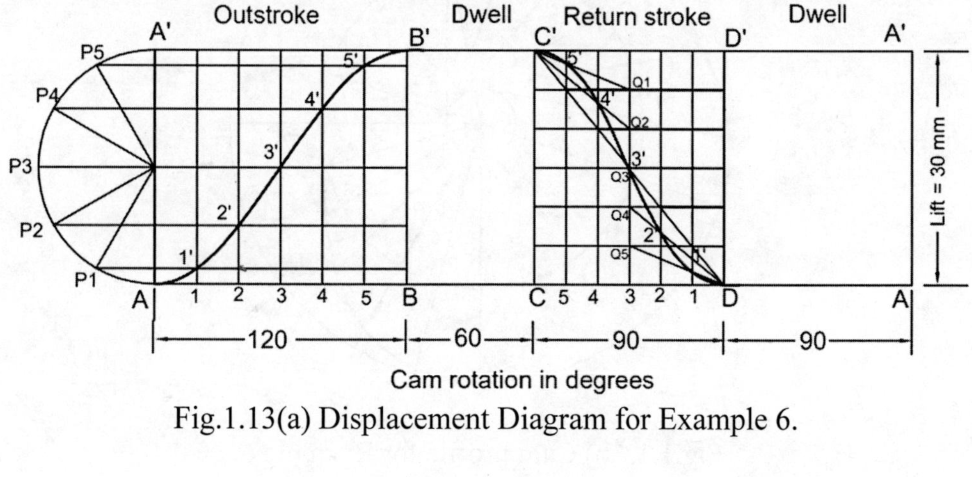

Fig.1.13(a) Displacement Diagram for Example 6.

Construction of Cam Profile: To sketch the cam profile, the following steps are used and are shown in Fig.1.13(b).

- Draw a base circle with a radius of OA=50 mm and a follower vertically above, as shown in the figure. Over the follower, draw a semicircle with a diameter of 20mm.
- Divide the semicircle into 6 equal parts and draw the line projected on diameter as described in the follower displacement diagram. Draw circles through these points from the cam centre.
- Divide the circle into four parts according to outstroke, dwell, return stroke and dwell from OA, as shown in the figure. Angle A'OB' = 120°, angle B'OC'=60° and angle C'OD'=90°. Further, subdivide the outstroke and return stroke sectors into six equal parts. Let points 1, 2, 3, 4, and 5 be on the base circle for outstroke, and 5, 4, 3, 2, and 1 for return stroke. Draw the radial lines of these points from the centre.
- Transfer the lengths, 1-1', 2-2' and so on from the displacement diagram for SHM outstroke and constant acceleration and deacceleration motion for return stroke over the base circle.
- Join the points A, 1', 2', 3',4', 5' and B' for out stroke and C', 5', 4', 3', 2', 1', and D for return stroke with a smooth curve. Dwell portions B' to C' and D to A are the arcs, as shown in the figure. This is the profile of the cam required.

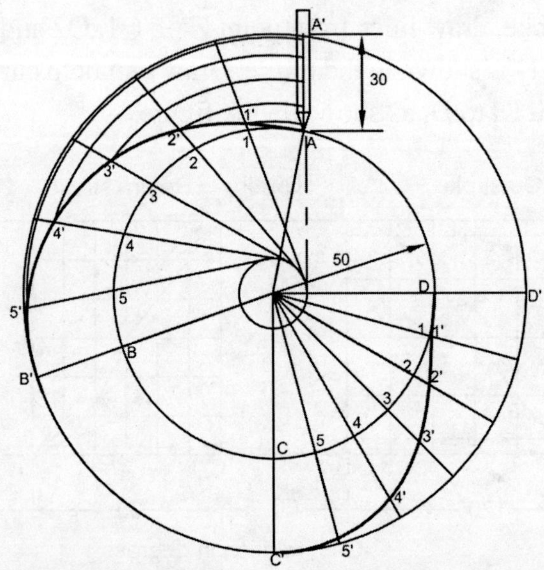

Fig.1.13(b) Cam profile for Example 6.

Example 7: A camshaft rotating at 1000 rpm drives a follower. It has an outward stroke of 40 mm during 100° rotation of the cam, returns in the next 90° to its original potion, and remains to dwell for the rest. Determine the maximum velocity and acceleration of the follower when it follows a constant acceleration and deceleration motion.

Solution

Given:

L = 40 mm = 0.04 m;

$\theta o = 100° \times \pi /180 = 1.745$ rad

$\theta_R = 90° = \pi /2 = 1.571$ rad; N =1000 rpm

The angular velocity of the camshaft $\omega = \frac{2\pi N}{60} = \frac{2\pi x 1000}{60} = 104.73$ rad/s

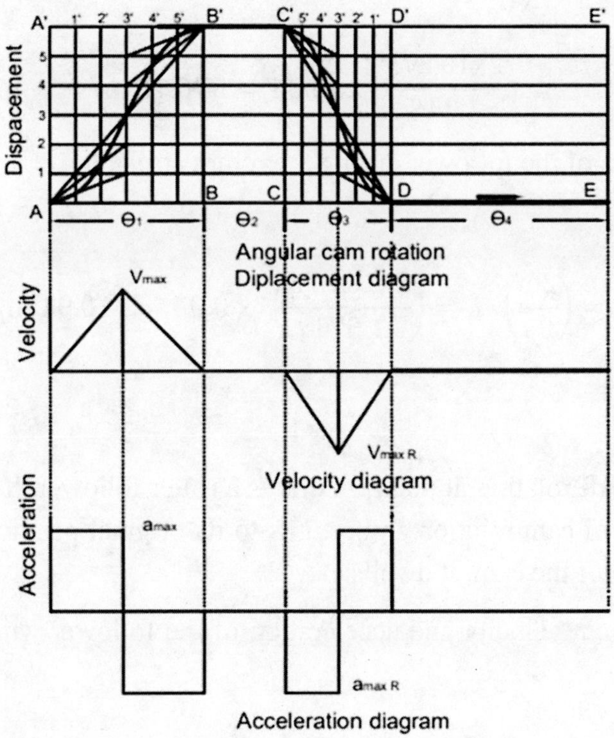

Fig.1.14 Diagrams for Example 7.

Referring to the figure Fig.1.14, the maximum velocity of the follower during the outstroke

$$v_o = \frac{2\omega L}{\theta_o} = \frac{2x104.73x0.04}{1.745} = 4.80 \text{ m/s}$$

The maximum velocity of the follower during the return stroke

$$v_R = \frac{2\omega L}{\theta_R} = \frac{2x104.73x0.04}{1.571} = 5.33 \text{ m/s}$$

Maximum acceleration of the follower during out stroke

$$a_o = \left(\frac{2\omega}{\theta_o}\right)^2 L$$

$$= \left(\frac{2x104.73}{1.745}\right)^2 x\ 0.04 = 576.16 \text{ m/s}^2$$

Maximum acceleration of the follower during the return stroke

$$a_R = \left(\frac{2\omega}{\theta_R}\right)^2 L = \left(\frac{2x104.73}{1.571}\right)^2 \text{x } 0.04 = 710.91 \text{ m/s}^2$$

Example 8: A camshaft rotating at 600 rpm drives a roller follower. It has an outward stroke of 20 mm during 75° of cam rotation and returns to its original position in the next 60°. For the remaining rotation of the cam, it dwells.

Determine the maximum velocity and acceleration of the follower when it follows a Simple Harmonic Motion.

Solution.

Given: L = 20 mm = 0.02 m, θ_o = 75 × π /180 = 1.31 rad, θ_R = 60 × π /180 = 1.04 rad,

N = 600 rpm

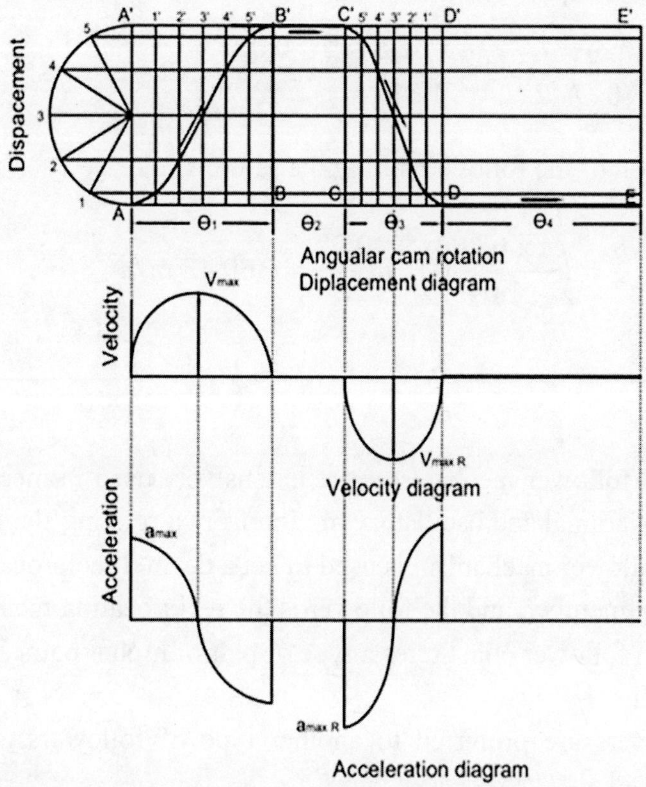

Fig.1.15 Diagrams for Example 8.

The angular velocity of the camshaft

$$\omega = \frac{2\pi N}{60} = \frac{2\pi \text{ x } 600}{60} = 62.84 \text{ rad/s}$$

Referring to the Fig.1.15, the maximum velocity of the follower during the outstroke

$$v_o = \frac{\pi\omega L}{2\theta_o} = \frac{\pi \text{ x } 62.84 \text{ x } 0.02}{2 \text{ x } 1.31} = 1.50 \text{ m/s}$$

The maximum velocity of the follower during the return stroke

$$v_R = \frac{\pi\omega L}{2\theta_R} = \frac{\pi \text{ x } 62.84 \text{ x } 0.02}{2 \text{ x } 1.04} = 1.89 \text{ m/s}$$

Maximum acceleration of the follower during out stroke

$$a_o = \left(\frac{\pi\omega}{\theta_o}\right)^2 \frac{L}{2} = \left(\frac{\pi \text{ x } 62.84}{1.31}\right)^2 x \frac{0.02}{2} = 227.17 \text{ m/s}^2$$

Maximum acceleration of the follower during the return stroke

$$a_R = \left(\frac{\pi\omega}{\theta_R}\right)^2 \frac{L}{2} = \left(\frac{\pi \text{ x } 62.84}{1.04}\right)^2 \text{x} \frac{0.02}{2} = 360.43 \text{ m/s}^2$$

Unit Summary

- The cam and follower mechanisms are extensively used in mechanical equipment to produce translational and oscillatory motion from a rotating shaft.
- A cam and follower mechanism is used to get a desired reciprocating motion. The cam is the rotating member, and the follower is the reciprocating member.
- A cam-driven follower constitutes a type of pair of higher pairs as the nature of contact is a line.
- Roller followers are preferred to another type of followers, as they offer smooth functioning and also less wear of cams.
- Radial cams are very popular among all types of cams. Other types are cylindrical cams and special cams for reciprocating or oscillating followers. The followers used along with cams are knife-edge followers, Roller followers, Spherical-faced followers, Flat-faced followers, etc.
- For slow and medium speed cams, uniform velocity, uniform acceleration and retardation, simple harmonic motion followers are used. Cycloidal and higher-order polynomial cams are recommended for high-speed cam follower applications like aero engines.
- The uniform velocity motion followers move linearly with the rotation of the cam, and a uniform velocity is achieved. The infinite acceleration or retardation of the follower is moderated by providing small curves on cam profiles. The maximum velocity is given by $V=L/(\theta/\omega)$.

- The follower with a simple harmonic motion gives a velocity diagram as a sine curve, while its acceleration diagram is a cosine curve. The maximum velocity and acceleration are given $v = \frac{\pi\omega L}{2\theta}$ and $a = \left(\frac{\pi\omega}{\theta}\right)^2 \left(\frac{L}{2}\right)$
- A follower moves with uniform acceleration and retardation. The displacement is a parabolic curve. This motion leads to a linear velocity and uniform acceleration while operating. The maximum velocity and acceleration are given by $v = \frac{2\omega L}{\theta}$ and $a = \left(\frac{2\omega}{\theta}\right)^2 L.$
- A cam-follower mechanism is provided with an offset to the follower to minimise the side thrust on the follower.

Multiple Choice Questions

1. The minimum displacement of the follower for a cam occurs on
   (a) prime circle, (b) pitch circle, (c) base circle, (d) pitch curve

2. The angle between the normal drawn to the pitch curve and the direction of the follower's motion is known as
   (a) pressure angle, (b) pitch angle, (c) base angle, (d) prime angle

3. A point on the cam profile at which the pressure angle is maximum is called
   (a) contact point, (b) pitch point, (c) centre point, (d) none of these

4. In automobile engines, the type of cam-follower generally used is
   (a) knife edge follower, (b) roller follower, (c) spherical-faced follower, (d) flat-faced follower

5. For air-craft engines, which type of cam follower is used?
   (a) knife edge follower (b) flat-faced follower (c) roller follower (d) spherical-faced follower

6. A radial follower is one that
   (a) oscillates, (b) reciprocates in the guides, and (c) translates along an axis of the cam. (d) swings

7. A cam-follower mechanism is provided with an offset to the follower to

   (a) minimise the side thrust, (b) minimize the jerk, (c) accelerate the motion, (d) increase the velocity

8. The point on the cam with the maximum pressure angle is known as
   (a) cam centre, (b) pitch point, (c) tracepoint, (d) prime point

9. The motion of the cam is transferred to the valves through
   (a) Pistons (b) Rocker arms (c) Camshaft pulley (d) Valve stems

10. A cam driving a roller follower constitutes a type of pair
    (a) lower pair, (b) higher pair, (c) close pair, (d) cam pair

11. For low and moderate-speed engines, the cam follower has a motion

    (a) Simple harmonic motion, (b) cycloidal motion, (c) uniform velocity, (d) uniform acceleration and retardation.

12 A cam follower with a simple harmonic motion has the acceleration at the end of the stroke is
(a) zero (b) maximum (c) maximum and minimum (d) maximum and zero

13 The maximum velocity of the cam follower, when it follows a Simple Harmonic Motion, is
(a) $v = \frac{\pi\omega L}{2\theta}$ (b) $v = \frac{\pi\omega L}{\theta}$ (c) $v = \frac{\pi\theta L}{2\omega}$ (d) $v = \frac{\pi\omega L}{4\theta}$

14 A cam follower moving with a constant acceleration and deacceleration motion has the velocity
(a) zero at the beginning and maximum at the end of the stroke.
(b) zero at the beginning and maximum at the middle of the stroke.
(c) constant through the stroke.
(d) maximum at the beginning and the end of the stroke.

15 The maximum velocity of the cam follower, when it follows a Simple Harmonic Motion, is

(a) $a = \left(\frac{\pi\omega}{\theta_o}\right)\frac{L}{2}$ (b) $v = \frac{\pi\omega L}{\theta_o}$ (c) $a = \left(\frac{\pi\omega}{\theta_o}\right)^2\frac{L}{2}$ (d) $a = \left(\frac{\pi\omega}{\theta_o}\right)^2 L$

Answers to Multiple Choice Questions

1. (c) 2. (a) 3. (b) 4. (c) 5. (c) 6. (b) 7. (a) 8. (b) 9. (b) 10. (b) 11(a)

12. (d) 13. (a) 14. (b) 15. (c)

Exercises

1. A disc cam revolving at a speed of 200 rpm drives a knife edge follower with uniform velocity. The follower-out stroke of 30 mm happens during the first half of the cam revolution. The follower returns to its original position with uniform velocity during the next 90° of rotation, and the remaining is the dwell period. The minimum radius of the cam is 60 mm. Calculate the velocity of the cam follower during the outstroke and return stroke. Draw the cam profile when the follower's axis passes through the camshaft's axis.

2. A cam rotating at a uniform speed of 300 rpm displaces a follower with simple harmonic motion. The radial or offset roller follower gives a maximum displacement of 25 mm during 120° of cam rotation. A constant displacement is observed during a dwell for 60° of cam rotation. Further, it comes to its initial position during the return stroke with 90° of cam rotation. The remaining 90° of cam rotation is again dwelled. The maximum radius of the cam is 20 mm, and the roller diameter is 8 mm.

   Draw the displacement diagram and cam profile when a) the follower's axis passes through the camshaft axis. b) the follower axis is offset by 20 from the camshaft axis.

   Calculate the maximum velocity and acceleration during the outstroke and return stroke for the above motion of the follower.

3. A cam rotating clockwise with a uniform speed drives a roller follower. The roller follower of 20 mm diameter has the following motion:

   (a) Outward through a distance of 25 mm during 100° of cam rotation with SHM

   (b) Dwell for 80° of cam rotation.

(c) Return to its initial position during 120° of cam rotation with uniform velocity and
(d) Dwell for the remaining cam rotation.

Draw the displacement diagram and cam profile.

4. A 30 mm minimum diameter camshaft has a uniform speed of 1200 rpm. The cam's minimum radius is 45 mm. At the same time, the follower is offset by 15 mm from the axis of the cam. Draw the displacement diagram and the cam profile. The cam drives the knife edge follower with equal uniform acceleration and retardation, which have the following motion:
   (a) Outward stroke of 25 mm for 120° of cam rotation and motion.
   (b) Dwell for 60° of cam rotation.
   (c) Return stroke during 90° of cam rotation.
   (d) Follower to dwell for the remaining 90° of cam rotation.

   Draw the displacement diagram and the cam profile for the cam-follower mechanism. Determine the follower's maximum velocity and uniform acceleration on both the outstroke and the return stroke.

5. The following data is available for a cam follower mechanism. The follower moves outward with simple harmonic motion while it moves with uniformly accelerated motion during return.
   The least radius of the cam = is 50 mm. The speed of the cam = is 360 rpm.
   Angle of outstroke = 48°, first dwell = 42°, return stroke = 60°.
   Lift of follower = 40 mm. Diameter of roller = 30 mm. Offset of cam = 20 mm.

   Find the maximum velocity and acceleration of the follower during function.

6. Draw a cam profile for a roller follower that moves with simple harmonic motion during the outstroke and while it travels uniformly accelerated and deaccelerated during the return stroke.

   Minimum cam radius = 50 mm.

   The angle of out stroke = 48°.

   Angle of dwell = 42°

   The angle of Return stroke = 60°

   Lift of follower = 40 mm

   Diameter of roller = 30 mm

Offset distance of follower with cam = 20 mm.

7. Draw a displacement diagram and cam profile for a cam operating with a knife edge follower having a lift of 30 mm. The follower follows an outstroke of 150° of its cam rotation with SHM. The follower keeps rest for a period of dwell for 60°.
The follower returns to its initial position for the 100° of the cam rotation with uniform velocity. This cam rotates with a uniform velocity of 120 rpm and has a least radius of 20mm. What will the following year's maximum velocity and acceleration during lift and return?

8. The cam-follower mechanism is used in printing press machinery. The link to raise with simple harmonic motion through 50 mm in 1/3 of a revolution during an outstroke. In this position, the follower keeps it fully raised during the 1/12 revolution. Then, it slowly comes to the initial position by harmonic motion in 1/6 revolution. The follower remains rested during the next part of the revolution. The follower is a roller and has a diameter of 20 mm. The minimum radius of the cam is 25 mm, and the diameter of the camshaft is 25 mm. The axis of the follower passes through the axis of the camshaft. If the camshaft rotates at a uniform speed of 100 rpm, find a follower's maximum velocity and acceleration during raise and lowering.

9. A cam rotating with a uniform speed drives a roller follower for a stroke of 25 mm. The diameter of the roller follower is 20 mm and has the following motion during operation.
   (a) Outward motion during 120° of cam rotation.
   (b) Dwell for 60° of cam rotation.
   (c) Return stroke during 90° of cam rotation and
   (d) Dwell for the remaining 90° of cam rotation.

   The minimum radius of the cam is 40 mm, and the follower's axis is offset by 10 mm from the axis of the cam. The motion of the follower follows simple harmonic motion on outward and uniform velocity during the return stroke. Draw the cam profile of the cam.

10. A cam rotating at a uniform speed is required to move an offset knife edge follower. The follower with a uniform and equal acceleration or retardation on both outstroke and return strokes is required. The angle of the outstroke, the angle of dwell (between

the end of the outstroke and the beginning of the return stroke) and the angle of the return stroke are 110°, 50° and 110°, respectively.

The follower dwells for the rest of the cam rotation. The minimum radius of the cam is 60 mm, while the lift of the follower is 30 mm. The diameter of the roller is 10 mm, and the follower is offset by 20 mm from the axis of the cam. Draw the displacement diagram and the cam profile.

Experiment:

Aim: 1) To plot the follower displacement v/s cam rotation for various cam-followers.
2) To determine the angle of the outstroke, return stroke and dwell periods.

CAM analysis apparatus:

This device consists of a rotating cam and a reciprocating follower. As the cam rotates, the follower reciprocates depending on the type of cam profile loaded. The relative position of the cam and follower can be assessed using this device. Cam is graduated with an angular position in degrees, and the follower displacement with a dial gauge.

(The cam analysis apparatus is provided with different sets of cams and followers)

Procedure:

- Fix the required cam and follower into the device.
- Adjust the position of the follower with cam, radial or offset set by a known distance.
- Set the cam position to zero and see that the follower is on the cam with a minimum radius.
- Fix the dial gauge on the top of the follower with preloaded sufficiently to ensure the dial gauge contacts the follower.
- Gradually rotate the cam and observe the follower moving out and returning. Rotate the cam a few rotations and get familiar with the mechanism.
- Note down the readings of the follower displacements and the corresponding angular positions of the cam. In Table 1.1
- While experimenting, observe the movement of the follower and note the following: Dwell (follower not moving), Beginning and end of outstroke and return. Enter the results in Table 1.2.
- Sketch the cam profile using the data and note the specification of the cam.

Observation:

Table 1.1 Observation Table.

Sl. No	Angular position (degree)	Follower displacement (mm)	Remarks (dwell/ rise/ return)

During the experiment, observe the movement of the follower and note the following. Dwell (follower not moving), Beginning and end of outstroke and return.

Table 1.2 Results.

Sl. No	Activity	Angular position (degree)		Duration (degree)
		Start	End	End-Start
1	Out stroke (Follower displacement increasing)			
2	Dwell-1			
3	Return stroke (Follower displacement decreasing)			
4	Dwell-2			

KNOW MORE

Lecture Series on Kinematics of Machines by Prof. Asok Kumar Mallik, Department of Mech. Engg. IIT Kanpur. Watch the NPTEL video on YouTube using the links:

https://www.youtube.com/watch?v=55tKVBVQDUY&t=103s
https://www.youtube.com/watch?v=oQrcPiQuCHI

Bibliography

- Theory of Machines, RS Khurmi and JK Gupta, S. Chand Publishing, 2005.
- Theory Of Machines, S. S. Rattan, McGraw Hill, 4th Edition, 2019.
- Theory of Machines and Mechanisms, John J. Uicker et al., Oxford University Press, Fifth Edition, 2017.
- Theory of Mechanisms and Machines, Amitabha Ghosh and Asok Kumar Mallik, East-West Press Private Limited, 1998.
- Theory Mechanisms and Machine, Jagdish Lal, Metropolitan Book Pvt Ltd., 1994

2 POWER TRANSMISSION

UNIT SPECIFICS

This unit presents the types, applications, advantages & limitations of various power transmission elements like belts, chains, ropes, and gears. These elements are very important in motion and power transmission in engineering. Simple calculations on belt tensions explore the maximum power transmission. The selection of chains and ropes, and their applications are very helpful in power transmission. Gear drives, types of gears, and gear trains for different applications are presented in this unit. The layout of various simple, compound, reverted, and epicyclic gear trains helps in understanding the working of gear drives.

RATIONALE

The transmission of power in automobiles and machinery is important for efficient and successful functioning. The knowledge of drives for power transmission for short and long distances is essential for mechanical engineers in designing drives.

PRE-REQUISITE

Nil

UNIT OUTCOMES

The list of outcomes of this unit is as follows:

U2-O1: Understand the role and application of power transmission.

U2-O2: Familiarise with belts, chains and ropes power transmission & capacity.

U2-O3: To understand gears, simple, compound and epicyclic gear trains.

Unit Outcomes	**Expected Mapping with the Course Outcomes** (2- Weak Correlation; 2- Medium Correlation; 3- Strong Correlation)				
	CO-1	CO-2	CO-3	CO-4	CO-5
U2-O1	3	3	3	1	1
U2-O2	3	3	3	1	1
U2-O3	3	3	3	1	1

2.1 Types of Drives

The motion and power must be transmitted from the driving shaft to the other driven shafts in the machinery. Various elements are employed to transmit power, such as belts, chains, gears and many. The flexible power transmission members like belts and ropes are used where the distance between the two shafts is moderate and large. The chains are also used for intermediate distances. While gears are used for power and motion transmission when the shafts are very close. The belts and rope drives use the friction between the belt or rope and the pulleys to transmit power. The slip and creep are common in flexible drives, and therefore, these drives are not positive drives. Gear and chain drives are called as positive drives because there is no slip. These drives are discussed in detail in the following sections.

2.2. Belt Drives and Applications

The belts are used to transmit power from one shaft to another shaft by means of pulleys. The pulleys are mounted on the shafts, and belts run over the surface of these pulleys. The pulleys are connected by an endless belt or rope passing over these pulleys and are shown in Fig.2.1. The connecting belt is kept in tension so that the motion of one pulley transmits power and motion to the other with minimal slip. The speed of the driven shaft can be varied using different diameters of the pulleys. A belt may be a rectangular section known as a flat belt or a trapezoidal section known as a V-belt, circular cross-section belts are used to transmit power.

1. Flat belt: These types of belts are common in factories and workshops, where a moderate amount of power is to be transmitted. The flat belt drive is preferred when the distance from one pulley to another is less than 8 metres.
2. V-belt: These types of belts are mostly used in industries where a moderate amount of power is transmitted for a shorter length. The major advantage of V-belts is that multiple belts can used in parallel to transmit more power.
3. Circular belt or rope: The circular belt or rope is mostly used in factories and workshops whenever a greater amount of power is transmitted from one pulley to another when the two pulleys are more than 8 meters apart.

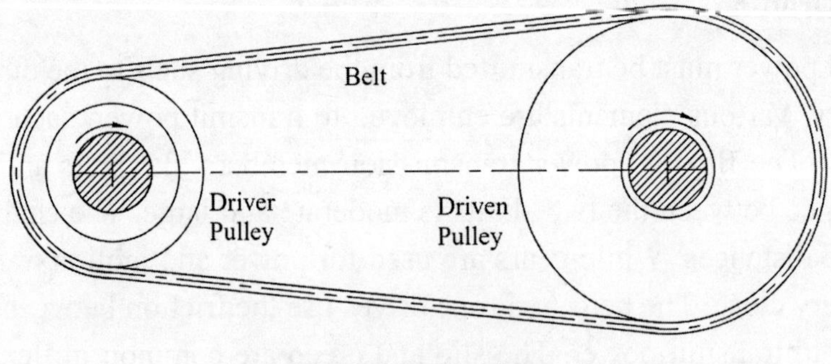

Fig.2.1 Belt drive.

Different types of belts are used in industries and at many places to transmit power from the driver shaft to other shafts within the equipment. Applications of belt and rope drives depend on speed, power and the distance between shafts. Light belt drives are those whose belt speeds are up to about 10 m/s and are common in agricultural machines and small machine tools. Medium drives are used to transmit medium power at belt speeds above 10 m/s but up to 22 m/s and are used in machine tools. Heavy drives are used to transmit large powers at belt speeds above 22 m/s, as in compressors and generators. Fig.2.2 shows the section of flat belts, V-belts and rope belts over the pulleys.

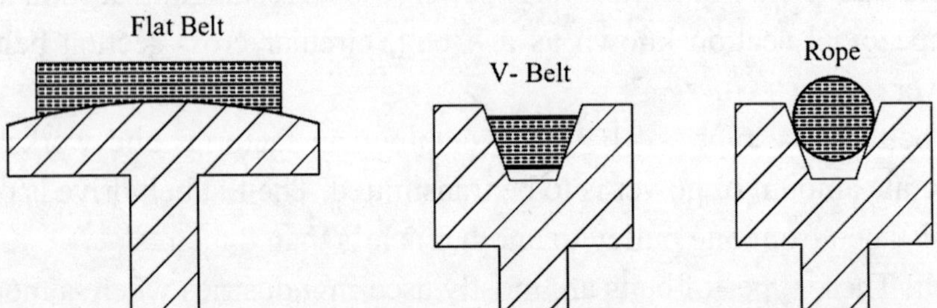

Fig.2.2 Sections of Flat, V-belt and Rope.

2.3 Material for flat and V-belts

The material used for flat or V-belts and rope drives is flexible, as they have to run over pulleys. The power is transmitted by the friction between the pulley and the belt; therefore, it must have a high coefficient of friction. Further, the belt material should be strong and durable. The following are commonly used belt materials.

Leather belts: The most popular material used for belts in the early days is leather. The long strips of leather are used to produce belts and usually are cut from the bulk. The hair side of a belt is in contact with the pulley surface, and there is more intimate contact between the belt and the pulley by the hair-side leather. The leather is oak-tanned or mineral salt-tanned. To increase the thickness of the belt, the strips are cemented together according to the number of layers: single, double or triple ply.

Cotton or fabric belts: The fabric belts are made of multiple layers of canvas or cotton and are stitched together. These belts are woven to the desired width and thickness. They are soaked with linseed oil to make the belts waterproof and prevent damage to the fibres. The cotton belts are cheaper, easy to manufacture and to join the ends. These belts are suitable for warm climates and exposed environments.

Rubber belts: Rubber belts are made of reinforced fabrics made of rubber. These belts cast and are endless. A thin layer of rubber on the faces keeps the belt and pulley less-slip. These belts are used where less heat is produced during power transmission. These belts are used where they are exposed to moisture. Balata belts are prepared from balata gum, which is similar to rubber. The good properties of balata are that the belts are acid-proof and waterproof. These belts are used in environments exposed to animal oils or alkalis. The balata belts become sticky at temperatures above 40°C. Comparatively, the strength of balata belts is 25% higher than that of rubber belts.

2.4 Angle of lap and Belt length

An open belt drive used to connect the shafts for power transmission is shown in Fig.2.3. The endless belt runs over the pulleys and connects both shafts. The belt is in contact with the pulley for a length and covered by an angle AO_1E, which is greater than 180° for the pulley at O_1. The angle of contact between the belt and the pulley is called the "Angle of lap". The belt drive transmits more power with the increased angle of the lap.

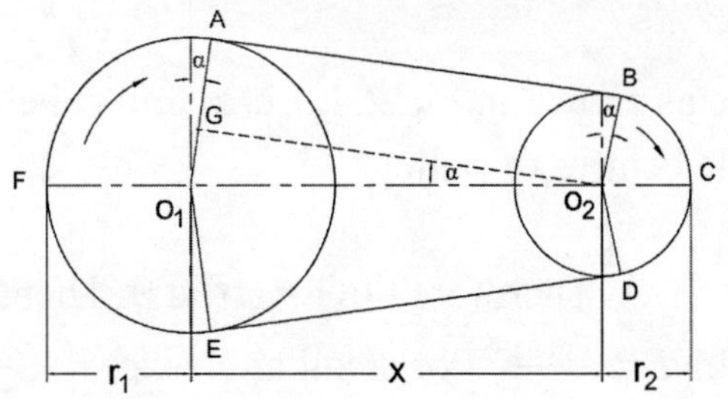

Fig.2.3. Length of open belt drive.

Open belt drive:

Consider an open belt drive as shown in Fig.2.3. Let r_1 and r_2 be the radius of pulleys and x be the distance between the centres of pulleys.

Draw a line O_2G parallel to AB. Therefore, $O_1G = (r_1 - r_2)$.

Then, from the triangle O_1GO_2

$$\sin\alpha = \frac{r_1 - r_2}{x} \quad \text{or for small angles, } \alpha = \frac{r_1 - r_2}{x}$$

The angle of lap or contact is $\theta = \pi r_1 + 2r_1\alpha$ on pulley with enter O_1.
For the other pulley, the angle of lap is $\theta = \pi r_2 - 2r_2\alpha$.

The length of the belt for an open belt drive is given by

$$L = Arc(EFA) + AB + Arc(BCD) + ED$$

$$L = \pi r_1 + 2r_1\alpha + x\, cos\alpha + \pi r_2 - 2r_2\alpha + x\, cos\alpha$$

$$L = \pi(r_1 + r_2) + 2\alpha(r_1 - r_2) + 2x\, cos\alpha$$

Where $\cos\alpha = \sqrt{1 - sin^2\alpha} = \sqrt{1 - \left(\frac{r_1 - r_2}{x}\right)^2}$

expanding by binomial theorem, we get $\cos\alpha = 1 - \frac{1}{2}\frac{(r_1 - r_2)^2}{x^2}$

substituting the above in Eqn. (1)

$$L = \pi(r_1 + r_2) + 2\frac{r_1 - r_2}{x}(r_1 - r_2) + 2x\left(1 - \frac{1}{2}\frac{(r_1 - r_2)^2}{x^2}\right)$$

$$L = \pi(r_1 + r_2) + 2\frac{(r_1 - r_2)^2}{x} + 2x - \frac{(r_1 - r_2)^2}{x}$$

The total length of the open belt drive is given by

$$\mathbf{L = \pi(r_1 + r_2) + 2x + \frac{(r_1 - r_2)^2}{x}} \qquad (1)$$

Cross belt drive:

Consider a cross belt drive as shown in Fig.2.4. Let r_1 and r_2 be the radius of pulleys and x be the distance between the centres of pulleys.

Draw a line O_2G parallel to DA. Therefore $O_1G = (r_1 + r_2)$. Then, from the triangle $O_1G\,O_2$

$$\sin\alpha = \frac{r_1 + r_2}{x} \quad \text{or small angles, } \alpha = \frac{r_1 + r_2}{x}$$

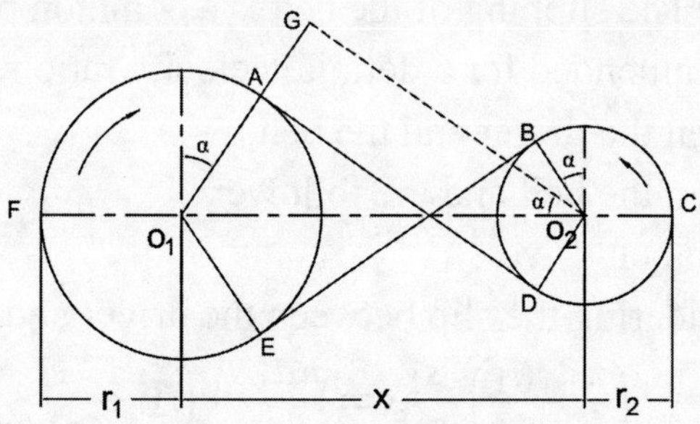

Fig.2.4. Length of cross belt drive.

The length of belt for cross belt drive is given by

$$L = Arc(EFA) + AD + Arc(DCB) + BE$$

$$L = \pi r_1 + 2r_1\alpha + x\cos\alpha + \pi r_2 + 2r_2\alpha + x\cos\alpha$$

$$L = \pi(r_1+r_2) + 2\alpha(r_1+r_2) + 2x\cos\alpha$$

$$\cos\alpha = \sqrt{1-\sin^2\alpha} = \sqrt{1-\left(\frac{r_1+r_2}{x}\right)^2}$$

by expanding by binomial theorem, we get $\cos\alpha = 1 - \frac{1}{2}\frac{(r_1+r_2)^2}{x^2}$

substituting above in Eqn. (1)

$$L = \pi(r_1+r_2) + 2\frac{r_1+r_2}{x}(r_1+r_2) + 2x\left(1-\frac{1}{2}\frac{(r_1+r_2)^2}{x^2}\right)$$

$$L = \pi(r_1+r_2) + 2\frac{(r_1+r_2)^2}{x} + 2x - \frac{(r_1+r_2)^2}{x}$$

The total length of the cross belt drive is given by

$$\mathbf{L = \pi(r_1+r_2) + 2x + \frac{(r_1+r_2)^2}{x}} \qquad (2)$$

The expression for the length of a belt of open or crossed is a function of r_1, r_2 and the centre of pulleys distance. Therefore, the length of the belt remains constant. The angle of lap is more in cross belt drive compared to open belt drive.

2.5 Slip and Creep

Slips in belts

In belt drive, the speed of the pulley is lesser than that of the belt because of the less frictional grip between the belt and the pulley surface. This leads to lesser belt speed than the driver pulley, and also, the belt moves without carrying the driven pulley with it. This is known as a "slip of the belt" and is expressed as a percentage. The slip reduces the speed ratio (N_1/N_2) of the drive. The slipping of the belt is a common phenomenon; therefore, the belt should never be recommended for a definite velocity ratio requirement.

Let s_1 % = Slip between the driver and the belt,

s_2 % = Slip between the belt and the follower.

s % = Total slip in belt drive

The velocity of belt, considering the slip between the driver and the belt

$$v = \frac{\pi d_1 N_1}{60} - \left(\frac{\pi d_1 N_1}{60}\right)\frac{s_1}{100} = \frac{\pi d_1 N_1}{60}\left(1 - \frac{s_1}{100}\right) \qquad (3)$$

The velocity of the follower considering the slip between the belt and the follower

$$\frac{\pi d_2 N_2}{60} = v - v\frac{s_2}{100} = v\left(1 - \frac{s_2}{100}\right)$$

Substituting the velocity v into the above, we get

$$\frac{\pi d_2 N_2}{60} = \frac{\pi d_1 N_1}{60}\left(1 - \frac{s_1}{100}\right)\left(1 - \frac{s_2}{100}\right) \qquad (4)$$

or

$$\frac{N_2}{N_1} = \frac{d_1}{d_2}\left(1 - \frac{s_1}{100} - \frac{s_2}{100}\right)$$ neglecting the product of s_1 and s_2

$$\frac{N_2}{N_1} = \frac{d_1}{d_2}\left(1 - \frac{s}{100}\right)$$ considering total slip

Creep in belts

During power transmission by a belt, it undergoes tension in the portion belt, which is called the tight side and the remaining portion of the belt is called the slack side. When the belt passes over the pulley, from the slack side to the tight side, the belt extends and contracts. This happens in every rotation of the belt. Because of this, a change in the length of the belt led to a relative motion between the belt and the pulley surfaces. The relative motion of the belt is called as a "creep". The effect of creep is to reduce the speed slightly of the driven pulley.

Let N_1 and N_2 be the speeds of the driver and driven pulleys. d_1 and d_2 are the diameters of the driver and driven pulleys. σ_1 and σ_2 are the stresses on the tight and slack of the belt. Further, E be Young's modulus of belt material. Considering the creep of belt, the velocity ratio is given by

$$\frac{N_2}{N_1} = \frac{d_1}{d_2} \mathrm{X} \frac{E+\sqrt{\sigma_2}}{E+\sqrt{\sigma_1}} \qquad (5)$$

2.6 Velocity Ratio

The velocity ratio relates the velocities of the driver and the follower. It is the ratio of the speed of the follower to the speed of the driver.

Let d_1 = Diameter of the driver in m
d_2 = Diameter of the follower in m
N_1 = Speed of the driver in rpm
N_2 = Speed of the follower in rpm
t = thickness of the belt in m

The length of the belt that passes over the driver $= \pi d_1N_1$ per minute
Similarly, length of the belt that passes over the follower $= \pi d_2N_2$ per minute
As belt that passes over the driver and the follower, length of belt is same in one minute,

$$\pi d_1N_1 = \pi d_2N_2$$

therefore, Velocity ratio $\frac{N_2}{N_1} = \frac{d_1}{d_2}$

Velocity ratio $\frac{N_2}{N_1} = \frac{d_1+t}{d_2+t}$ If thickness of the belt (t) is considered.

2.7 Ratio of tight side and slack side tension

Flat belts:

A driven pulley rotating at a constant speed with the angle of contact θ is shown in Fig. 2.5. Let T_1 and T_2 be the tensions on the tight and slack side. μ is the coefficient of friction between the belt and pulley. A small portion of the belt PQ, subtending an angle δθ at the pulley centre is considered to be in equilibrium. This portion belt PQ is in equilibrium under the following forces,

- Tension T in the belt at point P and Tension (T + δ T) in the belt at point Q;
- Normal reaction R_N, and Frictional force, $F = \mu \times R_N$

Resolving forces horizontally, neglecting $\delta T \delta \theta$

$$R_N = (T + \delta T)\sin\frac{\delta\theta}{2} + T\,\sin\frac{\delta\theta}{2} = T\delta\theta \qquad (6)$$

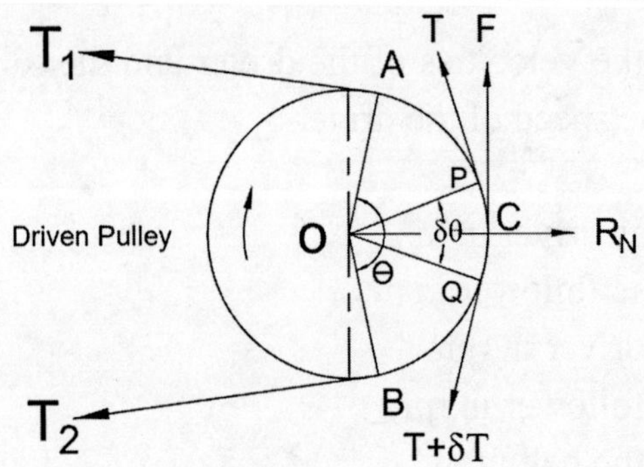

Fig.2.5 Ratio of belt tensions for a flat belt.

Resolving forces vertically

$$\mu R_N = (T + \delta T)\cos\frac{\delta\theta}{2} - T\cos\frac{\delta\theta}{2} = \delta T \qquad \text{considering } \cos\frac{\delta\theta}{2} \approx 1$$

Equating the values of R from above equations, we get

$$T\delta\theta = \frac{\delta T}{\mu} \quad \text{or} \quad \frac{\delta T}{T} = \mu\delta\theta \quad (7)$$

Integrating the Eqn. (7) both sides between the limits T_2 and T_1 and from 0 to θ respectively

$$\int_{T_1}^{T_2} \frac{\delta T}{T} = \int_0^{\theta} \mu\delta\theta$$

OR

$$\frac{T_1}{T_2} = e^{\mu\theta} \qquad (8)$$

This expression relates the tension of tight side and slack side tensions in terms of the coefficient of friction and the angle of contact.

V-Belts:

Consider a V-belt in a grooved pulley shown in Fig.2.6. The total normal reaction force R acting radially outward and the normal reaction between the belt and belt sides are in equilibrium.

Resolving the reactions vertically, we get

$$R = R_1 sin\beta + R_2 sin\beta = 2R_1 sin\beta$$

$$\text{Therefore, } R_1 = \frac{R}{2sin\beta}$$

$$\text{The frictional force} = 2\mu R_1 = 2\mu \frac{R}{2sin\beta} = \mu R cosec\beta$$

Consider a small portion of the belt covering an angle δθ at the centre and on the other side T+δT, similar to the previous section. Replace the frictional resistance μR to μR cosecβ. The relation between T1 and T2 for the V-belt drive is given by

$$\frac{T_1}{T_2} = e^{\mu\theta . cosec\beta} \qquad (9)$$

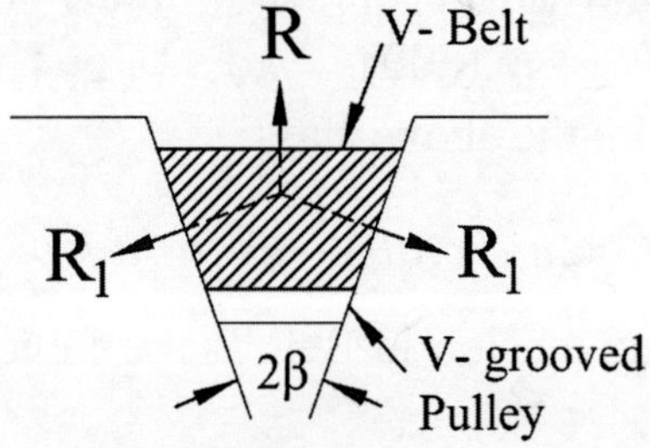

Fig.2.6 Reactions in a V-belt.

2.8 Centrifugal tension and initial tension

A centrifugal force is developed since the belt continuously runs over the pulleys at higher speeds. This force increases the tension on both the tight as well as slack sides. The tension caused by the centrifugal force is called centrifugal tension. At speeds lesser than 10 m/s, the magnitude of centrifugal tension is very small. However, the effect is considerable at belt speeds higher than 10 m/s, and the centrifugal forces should be considered in calculations.

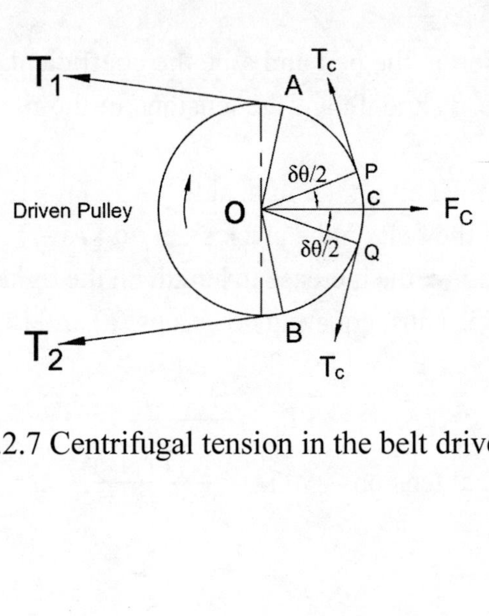

Fig.2.7 Centrifugal tension in the belt drive.

Consider a small sector PQ in the length of contact ACB. The Tensions T_c at points P and Q are tangent to the points at P and Q on the pulleys. The centrifugal force Fc acts radially outward from the centre O, as shown in Fig.2.7.

Let m = Mass of the belt per unit length,

v = Linear velocity of the belt,

r = Radius of the pulley, and

T_C = Centrifugal tension acting tangentially at P and Q.

The mass of belt of the sector PQ = m.(r.dθ)

Centrifugal force on the belt with the above mass

$$F_c = m.(r.d\theta)\frac{v^2}{r} = m.d\theta.v^2$$

The above centrifugal force and tension in Belt T_c are in equilibrium and equating force horizontally,

$$T_C \sin\left(\frac{d\theta}{2}\right) + T_C \sin\left(\frac{d\theta}{2}\right) = m.d\theta.v^2$$

Substituting $\sin\left(\frac{d\theta}{2}\right) = \frac{d\theta}{2}$ as the angle dθ is very small.

we get, $$2T_C\left(\frac{d\theta}{2}\right) = m.d\theta.v^2$$

or Centrifugal tension in belts $\boldsymbol{T_c = mv^2}$ (10)

Therefore, the total tension in the tight side, $T_{t1} = T_1 + T_C$ and

and total tension in the slack side, $T_{t2} = T_2 + T_C$

2.9 Initial tension in belts

Let T_0 be the initial tension in the belt and α be the coefficient of increase or change of the belt length per unit force. T_1 and T_2 are the tensions in the tight side and slack side of the belt.

Increase in the length of the belt on the tight side $= \alpha\,(T_1 - T_0)$ (a)

Decrease in the length of the belt on the slack side $= \alpha\,(T_0 - T_2)$ (b)

When it is at rest or in motion, the increase in length on the tight side is equal to the decrease in length on the slack side. Thus, equating equations (a) and (b),

$$\alpha\,(T_1 - T_0) = \alpha\,(T_0 - T_2)$$

$$T_0 = \frac{(T_1+T_2)}{2}$$

Considering the centrifugal tension $$T_0 = \frac{(T_1+T_2+2T_c)}{2} \quad (11)$$

2.10 Condition for maximum power transmission

Power transmitted by a belt drive:

Consider a belt drive used for the transmission of power from one shaft to another. Let T_1 and T_2 be the tensions in the tight side and slack side, respectively, and v be the velocity of the belt. The effective turning force at the circumference of the follower is the difference between the tensions ($T_1 - T_2$).

Therefore the work done per second $= (T_1 - T_2).v$

and power transmitted, $\boldsymbol{P = (T_1 - T_2).v}$ (12)

Condition for maximum power transmission:

We know that

$$\frac{T_1}{T_2} = e^{\mu\theta} \text{ or } T_2 = \frac{T_1}{e^{\mu\theta}}$$

Then, $$P = \left(T_1 - \frac{T_1}{e^{\mu\theta}}\right).v = T_1\left(1 - \frac{1}{e^{\mu\theta}}\right).v$$

Substituting $T_1 = T - T_c$ in above equation;

$$P = (T - T_c)\left(1 - \frac{1}{e^{\mu\theta}}\right).v$$

$$= (T - mv^2)\left(1 - \frac{1}{e^{\mu\theta}}\right).v$$

$$\mathrm{P} = (Tv - mv^3)\left(1 - \frac{1}{e^{\mu\theta}}\right) \qquad (13)$$

For maximum power, differentiate the above expression (13) with respect to v and equate to zero,

$$\frac{dP}{dv}\left[(Tv - mv^3)\left(1 - \frac{1}{e^{\mu\theta}}\right)\right] = 0$$

We obtain, $(T - 3mv^2) = 0$ or $T = 3T_c$

Therefore, the power transmitted is maximum when maximum tension is three times the centrifugal tension or

$$\boldsymbol{T_c = \frac{T}{3}} \qquad (14)$$

Example 1: A Belt drive is used to transmit power from a pulley 1.2 m diameter running at 240 rpm to a pulley 1.8 m diameter. Find the speed lost by the driven pulley as a result of the slip. The total slip is 0.5 % for the motion on both the pulleys.

Solution:

Given: $d_1 = 1.2$ m ; $d_2 = 1.8$ m; $N_1 = 240$ rpm

Speed of driver without slip $\mathrm{N_2 = N_1\frac{d_1}{d_2}}$ $\mathrm{N_2} = 240 \text{ x} \frac{1.2}{1.8} = 160$ rpm

Using slip equation for loss of speed due to slip

$$\frac{N_2}{N_1} = \frac{d_1}{d_2}\left(1 - \frac{s}{100}\right)$$

$$N_{2s} = 240 \text{ x } \frac{1.2}{1.8}\left(1 - \frac{0.5}{100}\right) = 159.2 \text{ rpm}$$

Loss of speed = N_2-N_{2s} = 160 - 159.2= 0.8 rpm

Example 2: A shaft of an engine is running at 200 rpm. is driving the machine shaft by means of a flat belt. The pulley on the engine shaft is of 1.2 m diameter and that of the machine shaft is 0.75 m diameter. The belt thickness is 5 mm. Calculate the speed of the machine shaft a) considering no slip and b) a total slip of 4%.

Given:

N_1 = 200 rpm; d_1 = 1.2 m; d_2 = 0.75 m; t = 5 mm = 0.005 m; s = 4%

Using equation;

$$\frac{N_2}{N_1} = \frac{(d_1+t)}{(d_2+t)}\left(1 - \frac{s}{100}\right)$$

$$N_2 = N_1 \frac{(d_1+t)}{(d_2+t)}\left(1 - \frac{s}{100}\right)$$

a) when there is no slip (s = 0)

$$N_2 = 200\, x\, \frac{1.2+0.005}{0.75+0.005}\left(1 - \frac{0}{100}\right) = 319.2 \text{ rpm}$$

a) when there is slip (s = 4%)

$$N_{2s} = 200\, x\, \frac{1.2+0.005}{0.75+0.005}\left(1 - \frac{4}{100}\right) = 306.43 \text{ rpm}$$

Example 3: A belt drive is used to transmit power from a driver pulley running at 200 rpm. The diameter of the driver pulley is 1 m, and the other pulley is 1.5 m in diameter. Determine the loss of speed by the driven pulley as a result of creep. The stresses on the tight and slack sides of the belts are 1.5 MPa and 0.5 MPa, respectively. Young's modulus of the material is 100 MPa.

Solution.

Given: d_1 = 1 m ; d_2 = 1.5 m; N_1 = 200 rpm; σ_1 = 1.5 MPa; σ_2 = 0.5 MPa; E = 100 MPa

Speed of follower neglecting the slip

$N_1 . d_1 = N_2 . d_2$ or $N_2 = \frac{N_1 d_1}{d_2} = \frac{200\, x\, 1}{1.5} = 133.3$ rpm

The belt creep equation is $\frac{N_2}{N_1} = \frac{d_1}{d_2} \text{ x } \frac{E+\sqrt{\sigma_2}}{E+\sqrt{\sigma_1}}$

Therefore, $N_2 = N_1 \text{x} \frac{d_1}{d_2} \text{x} \frac{E+\sqrt{\sigma_2}}{E+\sqrt{\sigma_1}}$

then $N_2 = 200 \text{ x } \frac{1.0}{1.5} \text{x} \frac{(100+\sqrt{0.5})}{(100+\sqrt{1.5})} = 132.6 \text{ rpm}$

Therefore, the loss of speed is (133.3 -132.6) = 0.7 rpm

Example 4: A belt drive is used connect two parallel shafts with two pulleys of diameters 0.7 m and 1.2 m. The distance between the centre lines of these shafts is 3 m. Find the length of the belt required for open belt drive as well as cross belt drive in order to change the direction of the rotation of shafts.

Solution.

Given: d_1 = 0.7 m ; d_2 = 1.2 m; x = 3 m.

Length of belt required or Open belt drive:

$$L = \pi(r_1+r_2) + 2x + \frac{(r_1-r_2)^2}{x}$$

Substituting values in above eqn. $L = \pi(0.35 + 0.6) + 2\text{x}3 + \frac{(0.35-0.6)^2}{3}$

= 9 m

Length of belt required or Cross belt drive:

$$\text{L} = \pi(r_1+r_2) + 2x + \frac{(r_1+r_2)^2}{x}$$

Substituting values in above eqn. $L = \pi(0.35 + 0.6) + 2x3 + \frac{(0.35+0.6)^2}{3}$

= 9.48 m

Example 5: The power transmitted by a belt running over a pulley of 500 mm diameter at 360 rpm. The coefficient of friction between the belt and the pulley is 0.25, the angle of lap is 190° and the maximum tension in the belt is 2000 N. Determine the capacity of the belt that can transmit power.

Given: d_1 = 0.5 m; N_1=360 rpm; θ = 190° = 3.317 rad; μ = 0.25; T_1=2000 N

v = (π. d_1. N_1)/60 =(π x 0.5 x 360)/60 = 94.26 m/s

Considering the equation for the ratio of belt tensions, $\frac{T_1}{T_2} = e^{\mu\theta}$

$$\text{or } T_2 = \frac{T_1}{e^{\mu\theta}}$$

$$T_2 = \frac{2000}{e^{0.25x3.317}} = 872.9 \text{ N}$$

Then, power transmitted by the belt $P = (T_1 - T_2).v$

P = (2000-872.9)/.872.9

P = 106240.4 W or 10.6 kW

Example 6: Two parallel shafts carrying pulleys, 0.450 m diameter and the other 0.250 m diameter, are 1.75 m apart. these are connected by a crossed belt. Find the angle of contact between the belt and pulley, and the length of the belt required.
The maximum permissible tension in the belt is 1000 N, and the coefficient of friction for the belt and pulley is 0.25. Calculate the power transmitted by this belt when the larger pulley rotates at 200 rpm.
Given:
d_1 = 0.45 m; d_2 = 0.25 m; x = 1.75 m; N_1=200 rpm; $\mu = 0.25$; T_1=1000 N

Velocity of belt: v = (π. d_1. N_1)/60

= (π x 0.45 x 200)/60 = 4.713 m/s

Angle of contact $sin\alpha = \frac{r_1+r_2}{x} = \frac{0.45+0.25}{2\,x\,1.75}$; θ = 11.53°

θ = 180+2x11.53 = 203° OR θ = 3.59 rad

Length of belt $L = \pi(r_1+r_2) + 2x + \frac{(r_1+r_2)^2}{x}$

Substituting values in above eqn. $L = \pi(0.225 + 0.125) + 2x1.75 + \frac{(0.225+0.125)^2}{1.75}$

L = 4.67 m

Power transmitted $P = \left(T_1 - \frac{T_1}{e^{\mu\theta}}\right) v$

$P = \left(1000 - \frac{1000}{e^{0.25x3.59}}\right)$ x 4.713 = 279 W

= 2.79 kW

Example 7: The power is transmitted to a pulley by a flat belt. The angle of lap is 120°. The belt is 100 mm wide by 6 mm thick and density1000 kg/m^3. The coefficient of friction is 0.3 and the maximum Tension in the belt is not to exceed 1200 N. Find the greatest power which the belt can transmit and the corresponding speed of the belt

Solution:

Given: $\theta = 120° = 2.095$ rad; $b = 0.1$ m; $t=0.006$ m; $\rho = 1000$ kg/m^3; $\mu = 0.25$; $T = 1200$ N.

Mass of the belt per metre $m = bt\rho = 0.1 x 0.006 x 1000 = 0.6$ kg/m

Speed of the belt for greatest power $v = \sqrt{\frac{T}{3m}} = \sqrt{\frac{1200}{3x0.6}} = 25.82\ m/s$

Centrifugal tension $T_c = \frac{T}{3} = \frac{1200}{3} = 400\ N$

Therefore, Tension in tight side $T_1 = T - T_c = 1200-400 = 800$ N

Let T_2 be the tension in slack side

$$\frac{T_1}{T_2} = e^{\mu\theta}$$

or $$\frac{800}{T_2} = e^{0.25x2.095}$$

Therefore, $T_2 = 473.8$ N

Maximum power transmit by belt $P = (T_1 - T_2)\ v = (800-473.8) \times 25.82$

$= 9669$ W or 9.67 kW

Example 8: V-belt drive is used to transmit power. The size of V- belt is 25 mm deep and maximum width is 20 mm and the included angle of V-groove is 30°. If the mass of the belt is 0.5 kg per metre length and the allowable stress for the belt material is 1.5 MPa. Determine the maximum power transmitted when the angle of lap is 130°. Assume $\mu = 0.15$.

Solution:

Given: $d = 25$ mm; $w = 20$ mm; $2\beta = 30°$; $\theta = 130° = 2.27$ rad; $m = 0.5$ kg/m;
$\mu = 0.15$; $\sigma = 1.5$ MPa

Maximum tension in belt: $T = \sigma(w.d) = 1.5 \times 10^6 \times 0.02 \times 0.025 = 750$ N

The velocity of the belt for maximum power to be transmitted

$$v = \sqrt{\frac{T}{3m}} = \sqrt{\frac{750}{3x0.5}} = 22.36 \text{ m/s}$$

Using the relation for tensions, $\frac{T_1}{T_2} = e^{\mu\theta\ cosec\beta}$

$$\frac{T_1}{T_2} = e^{0.15 \text{ x } 2.27 \text{ x } cosec(15)}$$

or $$\frac{T_1}{T_2} = 1.42 \qquad (1)$$

Centrifugal tension $T_c = \frac{T}{3} = \frac{750}{3} = 250$ N

Tension on tight side $T_1 = T - T_c$

$= 750 - 250 = 500$ N

Substituting in above Eqn. (1) $\frac{250}{T_2} = 1.42$ or $T_2 = 175.7$ N

The Maximum power transmitted by the belt drive is

$$P = (T_1 - T_2)\, v = (500 - 175.5) \times 22.36$$

$$= 7255 \text{ W} \quad \text{or} \quad P = 7.255 \text{ kW}$$

2.11 Chain Drives

The belt drive may not provide a fixed speed ratio between the pulleys. This is due to the slip and creep of belt material and is not used where a correct angular relation is required between the shafts. The chain drive is a mechanical operating system that offers a positive drive and keeps the speed ratio constant. The chains are made of steel links, hinged together to provide flexibility for warping around the toothed driving and driven wheels called sprockets. The teeth of the sprocket provide an exact space to fit the chain. The sprocket wheels and the chain are constrained to move together, ensuring no slip and a perfect velocity ratio. The sprocket and the chain are steel and require proper lubrication during working. The lubricant in the chain drive prevents rusting and also reduces the wear.

These chain drives are used to transmit power without slipping when the distance between the centres of shafts is short. A bush roller chain is very common in power transmission. It consists of outer and inner plates connected with pins, bushes and rollers. A pin is used to connect links and secure the bush and rollers firmly. These rollers protect the sprocket wheel teeth against wear.

A chain drive has several applications, which are for the transmission of power and the lifting of loads. Also used to carry heavy materials and woodworking machinery. They are found in the transportation industry, agriculture machinery and material handling equipment.

A chain drive has many advantages; those are:

- Chain drives require less space and are more compact as compared to belt drives as the chain and sprocket are made of steel. Chain drives can be operated at a higher temperature, whereas a belt cannot operate at such temperatures.

- Chain drives can operate in wet conditions, and they can also withstand abrasive conditions. The fire hazards do not face severe problems.
- They offer a constant velocity ratio as there is no slip between the chain and sprocket. No initial tension is required in chain drives. Chain drives are used up to 3 m in distance between two shafts.
- Chain transmits higher power than belt drive and has greater efficiency. The chain drives are easy to install and low maintenance.
- Multiple shafts can be driven from a single chain drive and can be used in reversing drives.

A chain drive has a few disadvantages, which are listed below:

- Chain drives require frequent lubrication for smooth running and to avoid rusting problems.
- The installation or initial cost is higher; chain drives need housing or covering. It needs accurate mounting and careful maintenance.
- It is a little noisy, hence a problem with vibration. The velocity fluctuation is noticeable on smaller sprockets or longer chain links or when stretched.

2.12 Selection of Chain & Sprocket Wheels

While selecting the chain and sprockets, the following features should be specified

- Power to be transmitted and speeds of the driver and driven shafts
- Sizes, types of driver and driven machinery, including environmental conditions
- Centre distance and layout of the shafts.

Sprocket selection

- Select the desired number of teeth on the sprockets (not less than 17 teeth and not more than 114 teeth)
- Determine the speed ratio according to the requirement. Determine the number of teeth on the input and output sprockets.
- Any chain drive operates at higher speeds or with impulse loads; the number of teeth on a small sprocket should have at least 25 teeth.

Chain selection:

- The following operating conditions arc considered in the selection of chains. Sprockets are correctly aligned, and the chain is maintained in the correct adjustment.
- Chain drive with two sprockets on parallel horizontal shafts having a speed ratio of from 1:3 to 3:1. Small sprocket with 19 teeth and different chain lengths (will affect chain life) and clean and adequate lubrication throughout the chain's life.
- Expected life of 15,000 hours, operating temperature between –5 °C to 70 °C and uniform operation without overload, shocks or frequent starts.

2.13 Methods of lubrication

Chains and sprockets are metallic links, and therefore, a noisy situation is developed. To increase the life of the chain drives, it is essential to lubricate the drive satisfactorily. Also, another disadvantage of worn chain drives is stretching of length. A well-lubricated chain drive reduces the noise and cools the chain drive at higher speeds. The clearance between the pin link plate and the other roller link plate on the slack side of the chain needs to be filled with oil.

The lubricant oil creates a thin film on the pin and bush, which minimizes frictional losses and also lessen the wear of links, thus increasing the chain's service life. Only good-quality lubricants should be used to lubricate the transmission chains. High viscous or heavy oils or grease is suitable are not suitable for chains. The viscosity of the oil to be used will depend on the chain's speed and size and the ambient temperature.

- **Manual Lubrication**: It is the simplest method of lubrication, and oil is applied with an oil brush in the gap between the pin link and roller link on the slack side of the chain. It should be applied frequently or when necessary to prevent the bearing area of the chain from becoming dry. Turn off the power switch before lubricating or servicing chain systems of higher capacity.
- **Drip Lubrication**: In this method, a simple chain casing is used. The oil container at an elevated place supplies the oil to the chain system by drip feed. Each strand of the chain will ordinarily receive the required drops of oil in a minute, according to the chain speed and other requirements.
- **Oil Bath Lubrication**: The chain drive is covered by a leak-free casing. The part of the chain and sprocket is dipped in oil, and it lubricates in the running. The minimum depth of oil should be maintained, and it should be noted that the overfilling of oil will be adversely affected by the heat generated.

- **Lubrication by Slinger Disc**: A slinger disc is fitted with the chain drive, and the drive is covered by a leak-free oil casing. Oil is splashed on the chain while rotating at higher speeds. The circumferential speed should be at least 3.5 m/s. If the width of the chain or number of parallel chains is used, then slinger discs are attached on both sides. The chain should not pass through the oil sump.
- **Lubrication using a Pump**: The oil is circulated in the chain drive using a pump. It is called as forced lubrication. The system uses a leak-free casing. The amount of oil supplied to each hole is constant. The number of such holes required to supply depends on the number of chain strands used. In circulation, the oil is cooled and circulation is repeated.

2.14 Gear Drives

Many machines demand motion transmission with an exact velocity ratio from one element to another. Also, a definite velocity ratio is of great importance in measuring devices. This is possible with the use of gears or toothed wheels. The gear drive can be used when the distance between the shafts is very small. Gears have several projections on circumference called teeth to avoid slip in motion. Further, these projections will fit into the corresponding recesses on the periphery of the similar gear.

Compared to other drives, the advantage of the gear drive is that it transmits the exact velocity ratio. Gears transmit a large power with high efficiency and provide a reliable service. However, manufacturing gears is difficult, and errors in cutting teeth result in vibrations and noise during the operation.

2.15 Spur gear terminology

The following terms are used in this chapter. A few of them are illustrated in Fig. 2.8.

Pitch circle: It is an imaginary circle by which a pure rolling action would give the same motion as the actual gear.

Pitch point: It is a common point of contact between two pitch circles.

Pitch surface: It is the surface of the rolling discs, which the meshing gears have replaced at the pitch circle.

Pressure angle: It is the angle between the common normal to two gear teeth at the point of contact and the common tangent at the pitch point.

Addendum: It is the radial distance of a tooth from the pitch circle to the top of the tooth.

Dedendum: It is the radial distance of a tooth from the pitch circle to the bottom of the tooth.

Addendum circle: It is the circle drawn through the top of the teeth and is concentric with the pitch circle.

Dedendum circle: It is the circle drawn through the bottom of the teeth. It is also called a root circle.

Circular pitch: It is the distance measured on the circumference of the pitch circle from a point of one tooth to the corresponding point on the next tooth. Mathematically, Circular pitch, $p_c = \pi D/T$. where D = Pitch circle diameter and T = No. of teeth.

Diametral pitch: It is the ratio of the number of teeth to the pitch circle diameter in millimetres. Mathematically, Diametral pitch, $p_d = T/D$.

Module: It is the ratio of the pitch circle diameter to the number of teeth. It is usually denoted by m. Mathematically, Module, $m = D/T$

Clearance: It is the radial distance from the top of the tooth to the bottom of the tooth, in a meshing gear.

Total depth: It is the radial distance between the addendum and the dedendum circles of a gear. It is equal to the sum of the addendum and dedendum.

Working depth: It is the radial distance from the addendum circle to clearance circle

Tooth thickness: It is the width of the tooth measured along the pitch circle.

Tooth space: It is the width of space between the two adjacent teeth measured along the pitch circle.

Face of tooth: It is the surface of the gear tooth above the pitch surface.

Flank of tooth: It is the surface of the gear tooth below the pitch surface.

Top land: It is the surface of the top of the tooth.

Face width: It is the width of the gear tooth measured parallel to its axis

Pressure angle: It is the angle between the common normal to two gear teeth at the point of contact and the common tangent at the pitch point. The standard pressure angles are 14.5° and 20°.

Profile: It is the curve formed by the face and flank of the tooth

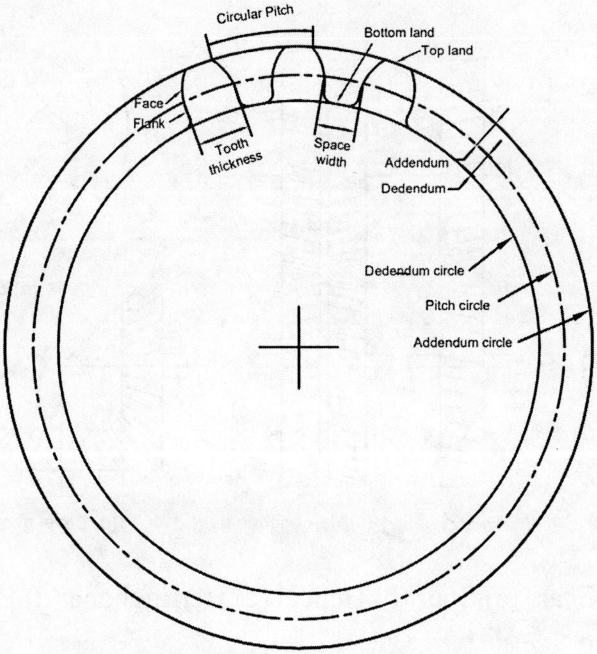

Fig.2.8 Common terms used in a gear.

2.16 Types of gears

The gears are used to transmit the motion and power between the shafts on which they are mounted. These gears are classified according to;

1. The position of the axes of two shafts

    (a) Parallel, (b) Intersecting, and (c) Non-intersecting and non-parallel.

The various gears are shown in Fig. 2.9. Two parallel and co-planar shafts are connected by the gears. These gears are *spur gears* in which teeth parallel to the axis of the gear (Fig.2.9(a)). Another type of gear may be *helical gear*, in which the teeth are inclined to the axis (Fig.2.9(b)). The double helical gears connecting parallel shafts are called as *herringbone gears* (Fig.2.9(c)). The two non-parallel or intersecting but coplanar shafts connected by gears are called *bevel gears* (Fig.2.9(d)). The two non-intersecting and non-parallel gears are called skew bevel or spiral gears (Fig.2.9(e)).

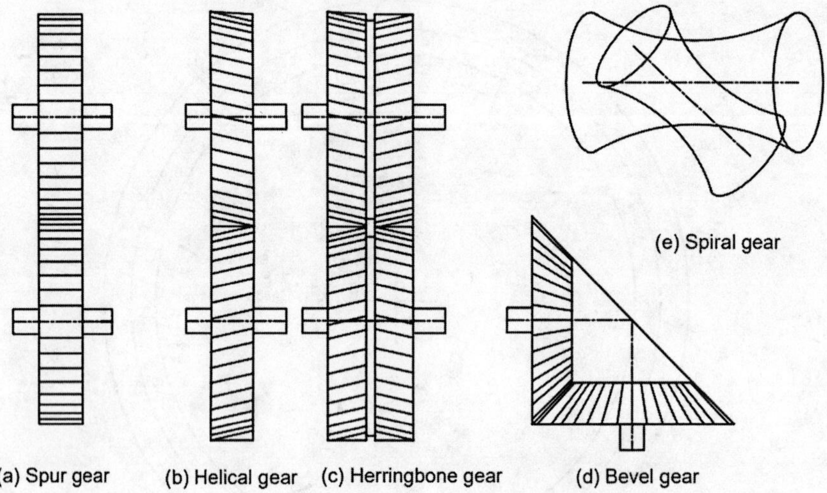

Fig.2.9 Types of Gears (a) Spur (b) Helical (c) Herringbone (d) Bevel and (e) Spiral.

2. The type of gearing.
   (a) External gearing,
   (b) Internal gearing, and
   (c) Rack and pinion.

In external gearing, the gears mesh externally with each other, as shown in Fig. 2.10(a). In external gearing, the motion of the two wheels is always unlike that they revolve in opposite directions.

For internal gearing, one gear has teeth cut on the inner side, and other shafts mesh internally with each other as shown in Fig. 2.10(b). The larger of these two wheels is called an annular wheel, and the smaller wheel is called a pinion. The motion of the two wheels is always like that they revolve in the same direction.

The gear teeth are cut in a straight line (radius of Gear in infinite) called a rack mesh externally with a pinion gear. Such type of gear is called *rack and pinion* and is shown in Fig. 2.10(c). The circular motion of the pinion is converted into the linear motion of the rack or vice-versa.

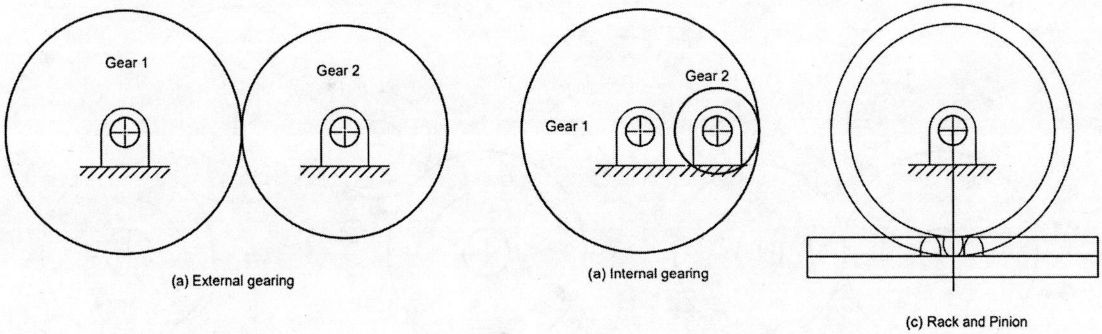

Fig.2.10 Types of gears (a) External (b) Internal and (c) Rack and pinion.

3. Type of teeth on the gear surface.
   (a) straight, (b) inclined, and (c) curved.

The teeth on a spur gear have straight teeth, whereas a helical gear has teeth inclined to the gear surface. In the case of spiral gears, the teeth are curved over the gear surface.

2.17 Types of Gear Trains

A combination of gears that are used for transmitting motion from one shaft to another is called a gear train. There are several types of gear trains depending upon the arrangement of gears and their axes. Fixed gear axes are found in a) Simple gear train, b) Compound gear train, and c) Reverted gear train, where gears revolve about axes. Epicyclic gear trains are those in which some axes are not fixed in space and are moving axes.

a) Simple gear train

All gears are in the same plane, and only one gear is on each shaft, as shown in Fig.2.11. These gears assembled to transfer motion are known as simple gear trains. In the figure, the gears are represented by their pitch circles. In a simple gear train, the number of gears depends on the distance between shafts and the direction of the final motion required. When the distance between the two shafts is small, the two gears, 1 and 2, are made to mesh with each other to form a simple gear train. The ideal gears are added in between if the distance is greater, as shown in Fig.2.11(b). Additional ideal gear may be added to transmit motion in the same direction, as shown in Figure.

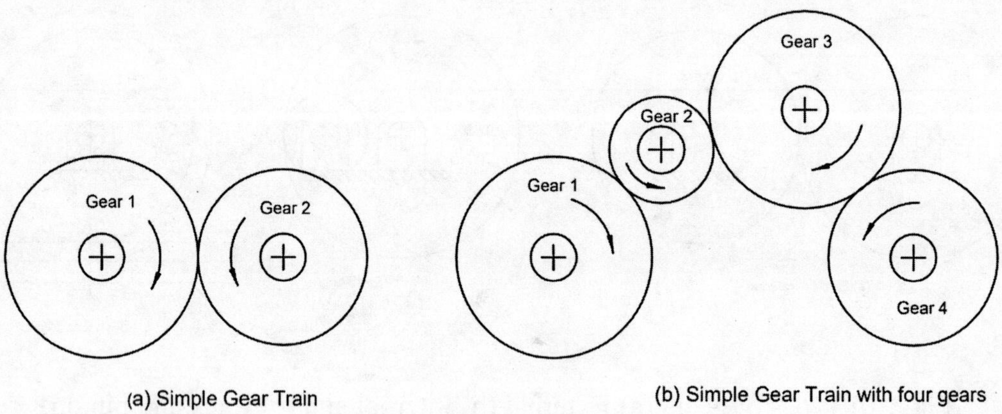

Fig.2.11 simple gear train.

The gear train's speed ratio (or velocity ratio) is the ratio of the speed of the driver to the speed of the driven.

Let N_1 = Speed of gear 1 in rpm,

N_2 = Speed of gear 2 in rpm,

d_1 and T_1 = Pitch diameter and number of teeth on gear 1,

d_2 and T_2 = Pitch diameter and number of teeth on gear 2.

m = module of tear tooth

The peripheral velocities of Gear 1 and Gear 2 are the same.

$$\pi d_1 N_1 = \pi d_2 N_2 \quad \text{or} \quad \pi(mT_1)N_1 = \pi(mT_2)N_2 \quad \text{since } d = mT$$

Therefore $$\frac{N_1}{N_2} = \frac{T_2}{T_1}$$

The ratio of the speed of the driven to the speed of the driver is known as *Train Value* of the gear train. Mathematically, Train value

$$\frac{N_2}{N_1} = \frac{T_1}{T_2}$$

In case of larger shaft distance, additional gears are used so that the speed ratio remains same (both magnitude and direction)

N_1, N_2, N_3 and N_4 be speeds of gear 1 to 4 respectively in rpm, and

T_1, T_2, T_3 and T_4 be teeth on gears 1 to 4, respectively.

The relation between speed and number of teeth for each set will be,

$$\frac{N_1}{N_2} = \frac{T_2}{T_1}, \quad \frac{N_2}{N_3} = \frac{T_3}{T_2}, \text{ and } \frac{N_3}{N_4} = \frac{T_4}{T_3}$$

Multiplying above equations, we get

$$\frac{N_1}{N_2} \times \frac{N_2}{N_3} \times \frac{N_3}{N_4} = \frac{T_2}{T_1} \times \frac{T_3}{T_2} \times \frac{T_4}{T_3}$$

or

$$\frac{N_1}{N_4} = \frac{T_4}{T_1} \quad (15)$$

Therefore, $speed\ ratio = \frac{speed\ of\ first\ driver}{speed\ of\ last\ follower}$

$$= \frac{number\ of\ teeth\ on\ last\ follower}{number\ of\ teeth\ on\ first\ Driver}$$

Similarly, the above equation holds good even if there are any number of intermediate or idle gears. These intermediate gears are called idle gears, as they do not affect the speed ratio or train value of the system. The idle gears are used when gear drive is required for a large centre distance and to get the desired direction of motion of the driven.

b) Compound gear train

A gear train with a shaft carrying more gears is called a compound gear train. In a simple gear train, idle gears do not affect the speed ratio of the system. Here, each intermediate shaft carries two gears fixed so that they rotate at the same speed. One of them meshes with the driver, and the other with the drive attached to the next shaft, as shown in Fig.2.12. The compound gear train is compact and saves space in machinery.

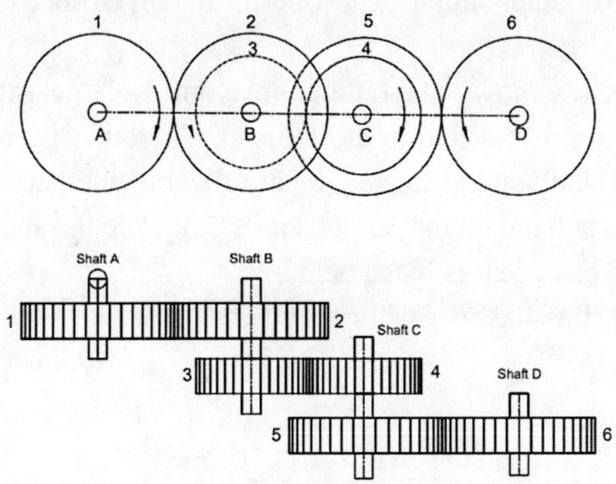

Fig.2.12 Compound gear train.

A compound gear train as shown in Fig.2.11. Shaft A is fitted with gear 1 and is the driving gear. This shaft A drives the shaft B that carries gears 2 and 3 (compounded gears). The gear 3 drives the shaft C through gear 4. Further, Shaft C carries compounded gears 4 and 5. Now gear 5 on shaft C drives the shaft D which is fitted with gear 6. In this compound gear train Gear 1 is the driving gear and gear 6 is the driven gear. All others are compounded gears.

Let T_1 = Number of teeth on driving gear 1

Similarly, T_2, T_2, T_3 ..., T_6 are teeth on gear 2 to gear 6 respectively

N_1 = Speed of gear 1

Similarly, N_2, N_3 ..., N_6 are speed of gear 2 to gear 6 respectively.

The speed ratio between the pair of gears is

$$\frac{N_1}{N_2} = \frac{T_2}{T_1}, \quad \frac{N_3}{N_4} = \frac{T_4}{T_3}, \text{ and } \frac{N_5}{N_6} = \frac{T_6}{T_5}$$

Multiplying the above equations

$$\frac{N_1}{N_2}\frac{N_3}{N_4}\frac{N_5}{N_6} = \frac{T_2}{T_1}\frac{T_4}{T_3}\frac{T_6}{T_5} \qquad (16)$$

As N_2= N_3 as they are on shaft B, and N_4 = N_5 as they are on shaft B, the equation (1) reduces to

$$\frac{N_1}{N_6} = \frac{T_2\,T_4\,T_6}{T_1\,T_3\,T_5} = \frac{\text{Product of number of teeth on the driven gears}}{\text{Product of number of teeth on the driver gears}}$$

The compound gear train provides a larger speed compared to the simple gear train. Usually, for a higher speed reduction, a compound train or worm gearing is used. The compound gear train is compact to another simple gear train for the given speed ratio.

Example 9: Fig.2.13 below shows a layout of a machine tool's gearbox. The motor shaft is connected directly to gear 1 and rotates at 600 rpm. The other gear wheels, 2 and 3, are mounted on a common shaft, and gears 4 and 5 are fixed to another shaft rotating together. The final driven gear 6, is fixed on the output shaft. Calculate the speed of gear 6.

The number of teeth on each gear is listed below:

T_1= 60, T_2= 50, T_3= 40, T_4= 30, T_5= 50, T_6= 40,

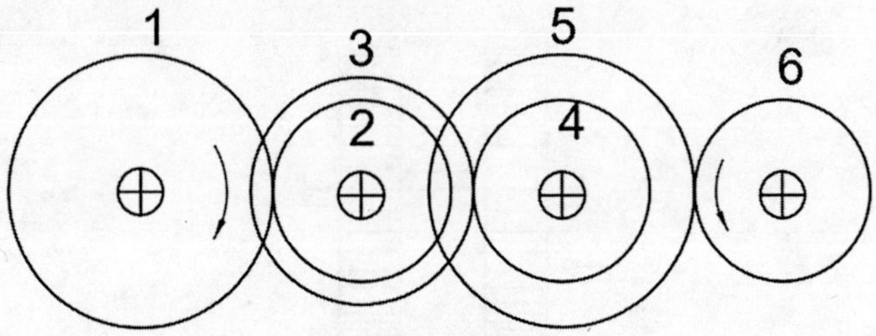

Fig.2.13 Layout of a machine tool's gearbox.

Given $N_1 = 200$ rpm,

Driving gears: 1, 3, and 5

Driven gears: 2, 4, and 6

We know the equation for gear ratio for compound gear train

$$\frac{N_1}{N_6} = \frac{T_2\, T_4\, T_6}{T_1\, T_3\, T_5} = \frac{\text{Product of number of teeth on the driven gears}}{\text{Product of number of teeth on the driver gears}}$$

Substituting the respective teeth in the above equation, we get

$$\frac{200}{N_6} = \frac{40 \text{ x } 30 \text{ x } 40}{60 \text{ x } 40 \text{ x } 50} = 0.4$$

Speed of final driven gear = 500 rpm

c) Reverted gear train

A reverted gear train is one in which the axes of the driver gear and the last driven gear are co-axial. The layout of the reverted gear train is shown in Fig.2.14. The first gear 1 drives gear 2 on the other shaft in the opposite direction. As gears 2 and 3 are mounted on the same shaft, gear 3 will rotate in the same direction as gear 2. The gear 3 drives the gear 4 in the same direction as that of gear 1. Therefore, a reverted gear train provides the motion of the first and last gear in the same direction.

Let T_1, T_2, T_3 and T_2 be the number of teeth on gears 1, 2, 3 and 4, respectively,
r_1, r_2, r_3. and r_4 be the pitch circle radius of the gears 1, 2, 3 and 4 respectively,
N_1, N_2, N_3 and N_4 be the speed of gears 1, 2, 3 and 4 respectively in rpm,

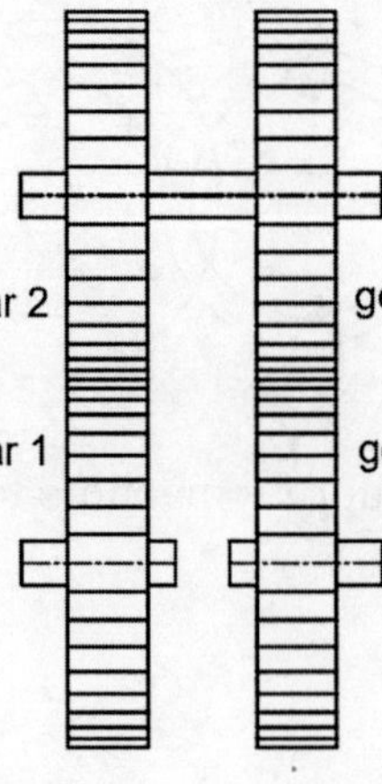

Fig.2.14 Reverted gear train.

The gear 1 drives the gear 2 in the opposite direction. As the gears 2 and 3 are fixed on the same shaft, they will rotate in the same direction as that of gear 2. The Gear 3 drives the Gear 4, and the direction is opposite to Gear 3; otherwise, it is the same as that of Gear 1. The distance between the centres of the shafts of gears 1 and 2, as well as gears 3 and 4, should be the same. Therefore,

$$r_1 + r_2 = r_3 + r_4$$

For gears, pitch circle radius = d/2 = module x No of teeth/2. Substituting for all the gears we get,

$$T_1 + T_2 = T_3 + T_4$$

$$\text{Gear ratio} = \frac{\text{Product of number of teeth on the driven gears}}{\text{Product of number of teeth on the driver gears}}$$

$$\frac{N_1}{N_4} = \frac{T_2 \text{ x } T_4}{T_1 \text{ x } T_3}$$

we can determine the number of teeth on each gear for the given centre distance, speed ratio and module only when the number of teeth on one gear is chosen arbitrarily. The reverted gear trains are used in automotive transmissions, lathe back gears, industrial speed reducers, and in clocks (where the minute and hour hand shafts are co-axial).

Example 10: A reverted gear train reduces the speed in the same direction of motion. The speed ratio of the reverted gear train is 16. The module for a set of gears A and B is 4 mm, and for gears C and D is 5 mm. Calculate the number of teeth for the gears. The minimum number of teeth on the gear should be more than 24. The distance between the centre of the shafts is 300 mm.

Solution:

Given: $N_A/N_D = 16$; $m_A = m_B = 4$ mm; $m_C = m_D = 5$ mm; $T_{min} = 24$.

$$\frac{N_A}{N_D} = \frac{N_A N_C}{N_B N_D} = 16; \quad \text{or} \quad \frac{N_A}{N_B} = \frac{N_C}{N_D} = \sqrt{16} = 4$$

$$\text{Or} \quad \frac{T_A}{T_B} = \frac{T_C}{T_D} = 0.25 \quad \text{as} \quad \frac{N_1}{N_2} = \frac{T_2}{T_1}$$

$$r_A + r_B = r_C + r_D = 300\,mm \quad or \quad \frac{m_A T_A}{2} + \frac{m_B T_B}{2} = 300$$

Therefore, we get by substituting the values of module

$$T_A + T_B = 150$$

Using Eqn (1) and (2), we get $T_A = 30,\ and\ T_B = 120$

Similarly, $r_C + r_D = 300\,mm$ and $\frac{T_C}{T_D} = 0.25$

$$\frac{m_C T_C}{2} + \frac{m_D T_B}{2} = 300 \quad \text{or} \quad T_C + T_D = 120$$

on solving, we get $T_C = 96,\ and\ T_D = 24$

Teeth on the gears are $T_A = 30,\ T_B = 120, T_C = 96,\ and\ T_D = 24$

2.18 Simple epicyclic gear train

An epicyclic gear train is one, in which the axes of the shafts on which the gears are mounted may move relative to the fixed axis. A simple epicyclic gear train is represented in Fig.2.15. The axis of gear A and the axis of arm C rotate about a common axis at O_A. The gear B rotates about its axis on the arm at O_B and meshes with gear A. If the arm is fixed, the gear train becomes simple and gear A can drive gear B or vice-versa. If gear A is fixed and then the arm is given a motion or rotated about the axis of gear A, then gear B is forced to rotate upon and around gear A. Such a type of motion is called as epicyclic.

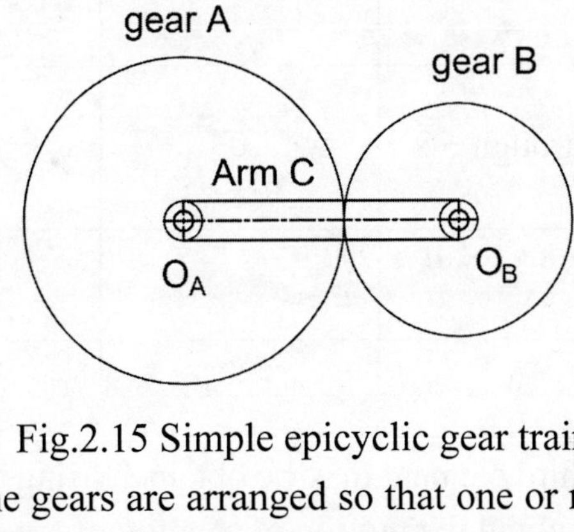

Fig.2.15 Simple epicyclic gear train.

In epicyclic gear trains, the gears are arranged so that one or more of their members move around another gear member. The epicyclic gear trains are classified as simple or compound. The applications of epicyclic gear trains are found in mechanical devices that demand high-velocity ratios. The main advantage of epicyclic gear is its compact size with moderate-sized gears. The Epicyclic gear trains are found in the back gear of lathes, differential gears used in automobile drives, hoists, pulley blocks, etc.

Velocity Ratio of Epicyclic Gear Train

Consider an epicyclic gear train, as shown in Fig.2.15. Let T_A and T_B be the number of teeth on gears A and B, respectively.

Step 1: Let us assume that the arm C is fixed. Then, it becomes a simple gear. When gear A makes one revolution anticlockwise (+ve direction), $N_B = -T_A/T_B$, Clockwise. [Note that $N_A/N_B = T_B/T_A$ and $N_B = 1$]. This statement of relative motion is entered in the first row of Table 2.1.

Step 2: If gear A makes +x revolutions, then gear B will make $N_B = -x(T_A/T_B)$ Clockwise. This statement is entered in the second row of the table.

Step 3: Finally, each element of an epicyclic train is given +y revolutions. Enter the values in the third row. The motion of each element of the gear train is added up and entered in the fourth row.

Revolution of epicyclic train output gear can be calculated from the last row of the Table.

Table 2.1. Motions of gear elements.

Step No	Conditions of motion	Revolutions of elements		
		Arm C	Gear A	Gear B
1	Arm C fixed. Gear A rotates through +1 revolution anticlockwise	0	+1	$-T_A/T_B$
2	Arm C fixed. Gear A rotates through + x revolutions	0	+ x	$-x(T_A/T_B)$
3	Add + y revolutions to all elements	y	Y	y
4	Total Motion	y	x+y	$y-x(T_A/T_B)$

The unknown speed of the third element may be obtained using the [y-x(TA/TB)] in the last row and any two conditions about the motion of rotation of the epicyclic gear train.

Example 11: An epicyclic gear train has two gears, A and B, on a rotating arm. Gear A and B have 40 and 60 teeth, respectively. Find the speed of gear B for the following cases.

a) Gear A is fixed, and the arm rotates at 200 rpm anticlockwise direction about the centre of gear A

b) Gear A makes 300 rpm in the clockwise direction,

Solution.

Given : $T_A = 40$; $T_B = 60$; $N_C = 200$ rpm. (anticlockwise)

The table of motions is prepared as per the configuration of the epicyclic gear train.

Table 2.2. Motions of gear elements.

Step No	Conditions of motion	Revolutions of elements		
		Arm C	Gear A	Gear B
1	Arm C fixed. Gear A rotates through +1 revolution anticlockwise	0	+1	$-T_A/T_B$
2	Arm C fixed. Gear A rotates through + x revolutions	0	+ x	$-x(T_A/T_B)$
3	Add + y revolutions to all elements	y	y	y
4	Total Motion	y	x+y	$y-x(T_A/T_B)$

a) The speed of Gear B when Gear A is fixed

When gear A is fixed, $x + y = 0$ from the above table

As the arm rotates at 200 rpm anticlockwise; $y = +200$ rpm.

Therefore, $x = -200$ rpm

The speed of Gear B is given by $[y-x(T_A/T_B)]$ from the above table

$$N_B = 200-(-200)(40/60)] = 333.3 \text{ rpm}$$

b) The speed of Gear B when the Gear and Arm are rotating

Gear A rotates at 300 rpm clockwise, $x + y = -300$

As the arm rotates at 200 rpm anticlockwise, $y = +200$ rpm

$$x + 200 = -300 \quad \text{or} \quad x = -500 \text{ rpm}$$

$\therefore$ Speed of gear B, $N_B = y-x(T_A/T_B)$

$$= 200-(-500)(40/60) = 533.3 \text{ rpm (anticlockwise)}$$

2.19 Methods of lubrication

Lubricating gears in machinery is essential as it increases the life of machinery and keeps a noise-free environment. It improves the efficiency of machinery because the energy loss due to friction is reduced. The purpose of lubricating gears is a) To reduce the sliding friction between teeth in gear trains. b) To reduce and transfer the heat caused by rolling and sliding friction of gear teeth.

There are three general lubrication methods for gearing systems

(1) Grease lubrication.

(2) Splash lubrication or oil bath method.

(3) Forced oil circulation lubrication.

The choice of lubricant and method largely depends upon the tangential speed of gears, rotating speed and the environment in which they work. Grease lubrication is a good choice at low speeds, up to 6 m/s. For medium and high speeds, say about 15 m/s, splash lubrication and forced oil circulation lubrication are appropriate. However, a grease lubricant is used for maintenance even with high speed.

Grease lubrication is preferred in low-speed / low-load applications. It is essential to see that greasing is done periodically, especially for gears of open-type usage. Since lubricants diminish or become depleted in open gear systems with use, periodic checks and refills are necessary. The proper selection of lubricant is essential. Usage of improper lubricants may cause damage to gear drives.

For gears at high speed / heavy load or special machines, the right type of lubricant, quantity, and methods according to the manufacturer's recommendation is important and be practised. The following are the three popular lubrication methods.

Grease Lubrication:

It is suitable for any gear system that is open or enclosed when the gear drive operates at lower speeds. When grease is used, the cooling effects are less than lubricating oil. This is a serious problem during temperature rise under high load and continuous operating conditions. A grease lubricant with good fluidity is effective in an enclosed gear drive system. A proper quantity of grease should be used in gear drives. Heavy grease loading can be harmful to the enclosed system. The excess quantity of grease may cause agitation and viscous drag, resulting in a power loss.

Splash Lubrication

Splash lubrication or oil bath method is preferred for use in an enclosed system. The rotating gears, including bearings, carry or splash lubricant onto the gear drive system. This method of lubrication needs a minimum of 3m/s tangential speed of gears. The concerns of splash lubrication are oil level and temperature limitations. The oil level during operation must be monitored. The excess agitation loss if the oil level is too high. Otherwise, there is no effective lubrication or ability to cool the gears if the level is too low. This problem can be tackled by the appropriate static level of lubricant in an oil pan.

The temperature of a gear drive system may lead to friction loss due to the relative motion of gears, bearings and also lubricant agitation. The rise in temperature is due to the lower viscosity of the lubricant or faster degradation of the lubricant. Other physical deformation of housing, gears and shafts mismatches and decreased backlash are also the reason for the temperature rise in gear drives.

Forced Oil Circulation Lubrication

In forced oil circulation lubrication, the contact portion of the teeth is lubricated continuously by means of an oil pump. Lubricants may be applied by any method of lubrication, like drop, spray, and oil mist methods. An oil pump circulates the lubricant and primarily drops or wets the contact portion of the gears. In an oil spray method, an oil pump is used to spray the lubricant on the contact area of the gears. Sometimes, lubricant mist is produced with compressed air to form an oil mist. It is sprayed against the contact region of the gears in high-speed gearing.

The method of forced lubrication is considered to be the best way to lubricate gears. This method is expensive as it requires an oil sump, pump, filter, piping and other devices are needed in the forced oil lubrication system. Therefore, it is used only for special high-speed gear drive applications.

2.20 Law of gearing

Consider a point of contact at Q on gear face when gear 1 meshes with gear 2 as shown in Fig.2.16. Let gears rotate in the directions as shown in the figure. The common tangent T-T and normal M-N are drawn to the curve at point Q. Draw O_1M and O_2N perpendicular to MN. During rotation, the point Q moves in the direction QC. Let v_1 and v_2 be the velocities of point Q on gears 1 and 2, respectively. If the teeth are to remain in contact, then the components of these velocities along the common normal MN must be equal

$$v_1 \cos\alpha = v_2 cos\beta$$

$$(\omega_1 \text{ x } O_1Q)\cos\alpha = (\omega_2 \text{ x } O_2Q)cos\beta$$

$$(\omega_1 \text{ x } O_1Q)\frac{O_1M}{O_1Q} = (\omega_2 \text{ x } O_2Q)\frac{O_2N}{O_2Q}$$

Or $$(\omega_1 \text{ x } O_1M) = (\omega_2 \text{ x } O_2N)$$

Also, from similar triangles O_1MP and O_2NP;

$$\frac{O_2N}{O_1M} = \frac{O_2P}{O_1P}$$

Therefore, $$\frac{\omega_1}{\omega_2} = \frac{O_2P}{O_1P}$$

Therefore, the angular velocity ratio is inversely proportional to the ratio of the distances of the point P from the centres O_1 and O_2.

A constant angular velocity ratio is obtained when the point P must be a fixed point (or pitch point for the two gears. In other words, *the common normal at the point of contact between a pair of teeth must always pass through the pitch point*. This is the fundamental condition is known as *Law of Gearing*.

Let d_1 and d_2 are pitch circle diameters of gear 1 and 2 having teeth T_1 and T_2 respectively. then velocity ratio is given by

$$\frac{\omega_1}{\omega_2} = \frac{O_2\,P}{O_1\,P} = \frac{d_2}{d_1} = \frac{T_2}{T_1}$$

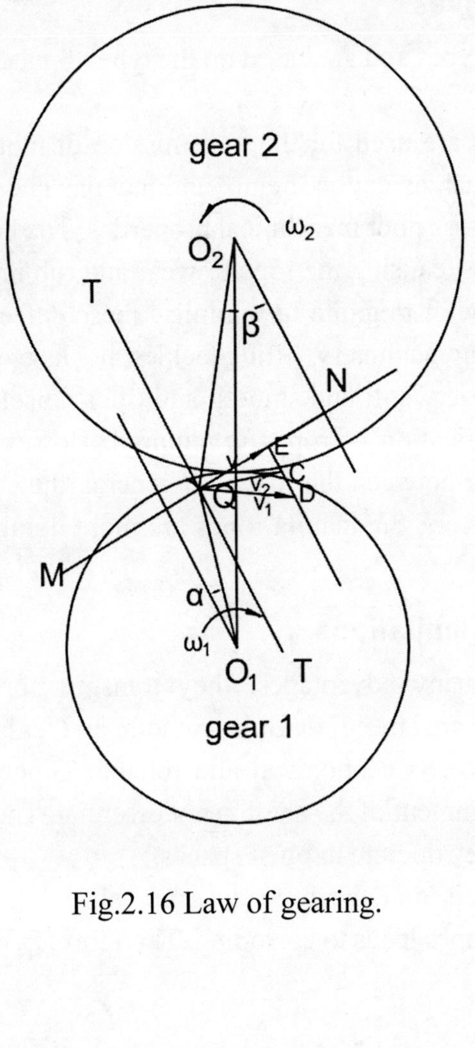

Fig.2.16 Law of gearing.

2.21 Rope Drives

The rope drives are widely used in lifts, mining shafts and material transportation where a large amount of power is transmitted from one pulley to another over a considerable long distance. The advantage of rope drive is that the frictional grip in rope drives is more than that in a V-drive since angle contact is greater. A rope drive connects a number of separate drives from the one driving pulley. The fibre rope drive can connect the pulleys up to 60 metres apart, and the wire ropes are used to drive pulleys up to 150 metres apart. The diameter of ropes usually ranges from 30 to 50 mm.

2.22 Types of rope drives

The rope drives use two types and are based on the type of material used. They are fibre and steel wires.

Fibre ropes: These ropes are used for the transmission of motion or power and are made from fibrous materials such as cotton, hemp, and manila. The hemp and manila fibres are not very flexible and possess poor mechanical properties. The hemp and manila ropes suffer from sliding over sheaves, causing the rope to wear and rub internally. The rope fibres are lubricated with tar, tallow or graphite to minimise these defects. These ropes are suitable for hand-operated hoisting machinery, lifting tackles, hooks etc.

Cotton ropes: These are very soft and smooth and offer noiseless drive. The advantage of this drive is that no lubrication of ropes is required. However, the lubrication of ropes reduces the external wear between the ropes. In general, the cotton ropes are costlier than manila ropes. Comparatively, the manila ropes are more durable and stronger than cotton ropes.

2.23 Advantages and limitations

The rope drives have many advantages: they transmit appreciable power for longer distances; the rope drives are strong, flexible, smooth, and less noisy. The drive operates in both directions. The drive is economical and reliable. Since the ropes run over longer distances, the precise alignment of the shaft is not essential. The rope drives are affected by outdoor conditions, as they operate in open spaces.

The major disadvantage is that the rope does not provide any sign in case of failure. Improper lubrication of ropes leads to corroding. Therefore, frequent inspection of the drive is required.

Ratio of Driving Tensions for Rope Drive

The ratio of driving tensions for the rope drive may be obtained in the similar way as V-belts. Refer the Fig. 2.17 for the forces acting on the rope. The ratio of driving tensions is given by

$$\frac{T_1}{T_2} = e^{\mu\theta\ cosec\beta}$$

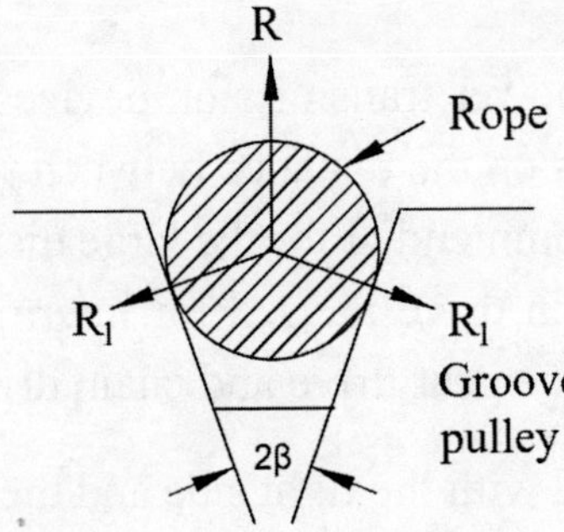

Fig.2.17 Forces on the rope.

Example 12: A rope drive is used transmit a power of 200 kW from a sheave of effective diameter 3 m, which runs at a speed of 100 rpm. The angle of lap is 180° ; the angle of groove 45° ; the coefficient of friction 0.28 ; the mass of rope 1.8 kg / m and the allowable tension in each rope 2000 N. Find the number of ropes required.

Solution:

Given: P = 200 kW, d = 3 m, N = 100 rpm, θ = 180° = 180 × π / 180 = 3.142 rad, 2 β = 45° or β = 22.5°, μ = 0.28, m = 1.8 kg/m, T = 2000 N.

Velocity of rope $v = \frac{\pi dN}{60} = \frac{\pi \text{ x } 3 \text{ x } 100}{60} = 15.71\, rad/s$

Centrifugal tension, $T_C = mv^2 = 1.8\,(15.71)^2 = 444.2$ N

and tension in the tight side of the rope, $T_1 = T - T_C = 2000 - 444.2 = 1555.8$ N

Tension in the slack side of the rope

$$\frac{T_1}{T_2} = e^{\mu\theta \, cosec\beta}$$

Substituting the values into the above equation, we get

$$\frac{1555.8}{T_2} = e^{0.28x\ 3.142x\ cosec(22.5)} \quad \text{or} \quad T_2 = 156.15 \text{ N}$$

The power transmitted per rope = ($T_1 - T_2$) v = (1555.8-156.15) x 15.17 = 21,232 W

Therefore, number ropes required to transmit a power of 200 kW

$$n = \frac{200}{21.232} = 9.42$$

Say 10 ropes are required to transmit the required power

Unit Summary:

The most common mechanical power transmission devices are belt, chain, rope and gear drives. The type of drive depends on the distance between the shafts, slip, and power to be transmitted. The belt drive is recommended for the large distance of shafts, and no effect of slip on power transmission. Chain drive is used for intermediate distances. Gear drive is used for short centre distances. The gear drive and chain drive are positive drives.

In belt drive, power is transmitted with the tight side and the other side slack side. The ratio of this tension depends on the angle of contact and coefficient of friction. The power transmitted in belts is decided based on the transmitted by the pulley on which the product of ($\mu\theta$) is small. Due to the considerable mass of the belt, centrifugal tension reduces the power transmitted. A belt drive transmits the maximum power at a speed when the centrifugal tension is one-third of the maximum possible tension. For shorter distances, V-belt drives are used.

The type of gear drive is decided by the layout of gear shafts. For parallel shafts, spur or helical gears are used, while bevel gears are used for intersecting shafts. For skew shafts or non-interesting shafts, worm and worm gears are used. When the distance between the shafts is larger and a positive drive is required, chain drives are used.

Multiple Choice Questions

1. For maximum power transmission by a belt drive, the ratio of centrifugal tension T_c to maximum tension T should be equal to

   (a) 2 (b) $\sqrt{2}$ (c) 3 (d) 1/3
2. In a cross-belt drive, the velocity ratio (N_1/N_2) of two pulleys connected is given by

   (a) D_1/D_2 (b) D_1D_2 (c) D_2/D_1 (d) D_1+D_2
3. The centrifugal tension T_c due to the mass of the belt, results in

   (a) increased power transmission (b) decreased power transmission (c) no effect on the power transmission (d) increased power transmission up to 1/3 of maximum velocity
4. The initial tension in the belt drive when stationary is

   (a) tension in the tight side (b) tension in the slack (c) sum of the tensions

(d) average tensions

5. In belt drive, the diameter of the driver pulley is half of the driven pulley. The angle of lap for the driven pulley is

   (a) equal to180° (b) less than 180° (c) more than 180° (d) more than 270°

6. Power transmitted by the flat belt is given by

   (a) $(T_1-T_2)/v$ (b) $(T_1+T_2)/v$ (c) $(T_1-T_2)/v^2$ (d) $(T_1-T_2)v$

7. A spur gear drive has a set of gears with 36 and 72 teeth. If both gears have 6 mm modules, the centre distance between the shafts is

   (a) 324 mm (b) 108 mm (c) 648 mm (d) 216 mm

8. The module is the reciprocal of

   (a) diametral pitch (b) radius of gear (c) pitch diameter (d) circular pitch

9. When two coplanar and parallel shafts are connected by gears having teeth inclined to the axis of the shafts, then the type of gearing is

   (a) spur (b) helical (c) bevel (d) spiral

10. The spiral gears are used to connect two non-parallel non-intersecting shafts are

    (a) parallel (b) intersecting (c) non-parallel and non-intersecting (d) parallel and non-intersecting

11. An imaginary circle on the gear provides a pure rolling action, which gives the same motion as the actual gear is

    (a) clearance circle (b) addendum circle (c) dedendum circle (d) pitch circle

12. A radial distance from the pitch circle to the top of a gear tooth is

    (a) dedendum (b) addendum (c) clearance (d) working depth

13. In spur gear, the module is the reciprocal of

    (a) diametral pitch (b) pitch diameter (c) circular pitch (d) addendum

14. The size of a gear is generally given by

    (a) circular pitch (b) pressure angle (c) diametral pitch (d) pitch circle diameter

15. The product of the circular pitch and the diametral pitch of a gear is equal to

    (a) π (b) $1/\pi$ (c) $1/2\ \pi$ (d) 2π

Answers to Multiple Choice Questions

1.(d), 2.(c), 3.(b), 4.(d), 5.(b), 6.(d), 7.(a), 8.(a), 9.(b), 10.(c), 11.(d), 12.(b), 13.(a), 14.(d), 15.(a),

Exercises:

1. Belt drive is used to transmit power from a driver pulley of 1.5 m diameter running at 300 rpm to a pulley 1.2 m diameter. Find the speed lost by the driven pulley as a result of the slip. The slip is 4 % for the motion on both the pulleys. [15 rpm].

2. A belt drive is used to transmit power from a driver pulley running at 240 rpm. The diameter of the driver pulley is 0.75m, and the other pulley is 1.5 m in diameter. Determine the speed of the driven pulley as a result of creep. The stresses on the tight and slack sides of the belts are 1.2 MPa and 0.7 MPa, respectively. Young's modulus of the material is 100 MPa. [119.45 rpm].

3. Two parallel shafts are connected by a belt drive with two pulleys of diameters 500 mm and 700 mm. The distance between the centre lines of these shafts is 4 m. Find the length of the belt required for open belt drive as well as cross belt in order to change the direction of the rotation of shafts. [9.88, 10 m].

4. A belt drive is used connect two parallel shafts with two pulleys of diameters 0.85 m and 1.5 m. The thickness of belt in 6mm. The distance between the centre lines of these shafts is 3.5 m. Find the length of the belt required considering the belt thickness for a) Open belt drive and b) cross belt drive in order to change the direction of the rotation of shafts. [9.74, 10.65 m].

5. The power transmitted by a belt running over a pulley of 600 mm diameter at 300 rpm. The coefficient of friction between the belt and the pulley is 0.24, angle of lap 160° and maximum tension in the belt is 2200 N. Determine the capacity of belt that can transmit power. [10.12 kW].

6. Two parallel shafts carrying pulleys, 0.60 m diameter and the other 0.40 m diameter are 2 m apart and are connected by a open belt drive. Find the angle of contact between the belt and each pulley and the length of the belt required. The maximum permissible tension in the belt is 1500 N, and the coefficient of friction for the belt and pulley is 0.25. Calculate the power transmitted by this belt when the larger pulley rotates at 240 rpm. [5.57 m, 6.28 kW].

7. A shaft delivers 6 kW at 200 rpm to another shaft rotating at 300 rpm through a open belt drive. The distance between the shafts is 4 m. The smaller pulley is 0.5 m in diameter. Calculate the Tensions in the belt. Take $\mu = 0.3$. [1267 N, 503 N].

8. A belt drive connects of two shafts 1.95m apart by a crossed belt. Diameters of the pulleys are 0.45 m and 0.2 m. Calculate the length of the belt required and the angle of lap between the belt and pulleys. Determine the power transmitted by the drive, if the speed of larger pulley is 200 rev/min. Assume the maximum permissible tension in the belt as 1 kN, and the coefficient of friction as 0.25.[4.975m, 199.2°, 2.74 kW].

9. Two parallel shafts carries pulleys connected by open belt drive. The diameters of pulleys are 1.5 m and 1 m and are 4.8 m apart. The smaller pulley rotates at 400 rpm. The initial tension in the belt is 3 kN and the mass of the belt is 1.5 kg / m length. The coefficient of friction of the belt drive is 0.3. Considering the centrifugal tension, calculate the power transmitted. [42.1 kW].

10. A compound gear train consists of 6 gears. The number of teeth on the gears A, B, C, D, E, and F are 60, 50 50, 40, 30 and 40. Gears B and C are on the second shaft, while gears D and E are on the third shaft. Gear A drives gear B, gear C drives gear D, and Gear E drives gear F. If gear A runs at 100 rpm, find the speed of gear F. 112.5 rpm].

11. A reverted gear train with a speed ratio of 12 is used to reduce the speed output shaft. The module for gears A and B is 3.125 mm, and for gears C and D are 2.5 mm. Calculate the suitable numbers of teeth for the gears. Minimum number of gear teeth is 24 teeth on any gear. [28, 100, 36, 124 and 12.3].

12. A rope drive is required to transmit 150 kW from a pulley of 1 m diameter running at 450 rpm. The safe pull in each rope shall not exceed 800 N, and the mass of the rope is 0.46 kg per metre in length. The angle of the lap and the groove angle are 160° and 45°, respectively. If the coefficient of friction between the rope and the pulley is 0.3, find the number of ropes required. [14].

Experiment:

Aim: To study the epicyclic gear train motion and to determine the epicyclic gear ratio.

Procedure:

The ratio of the speed of driver wheel to the speed of the driven wheel is called the speed ratio or velocity ratio. In a simple epicyclic gear train consists of two co–axial shafts. One sun gear (A), planetary gear (B) and a link (C) to connect and support gears, as shown in Fig.2.18.

- Set the required gears on the epicyclic gear train set up. Ensure that the motion of the epicyclic gear train is completely constrained.
- Note down the rotation of gears; driver and output shaft.
- Let T_A and T_B be the number of teeth on gears A and B, respectively.
- Let us fix the arm C. When gear A makes one revolution anticlockwise (+ve direction), $N_B = -T_A/T_B$, clockwise.
- If gear A makes +x revolutions, then gear B will make $N_B = -x(T_A/T_B)$ Clockwise. This statement is entered in the second row of the table.
- Finally, each element of an epicyclic train is given +y revolutions. Enter the values in the third row. The motion of each element of the gear train is added up and entered in the fourth row.
- Revolution of epicyclic train output gear can be calculated from the last row of the Table.
- Note down the angular rotation of arm and gear B.
- Calculate the epicyclic gear train.
- Compare the results obtained from the experimentally and from the calculations.

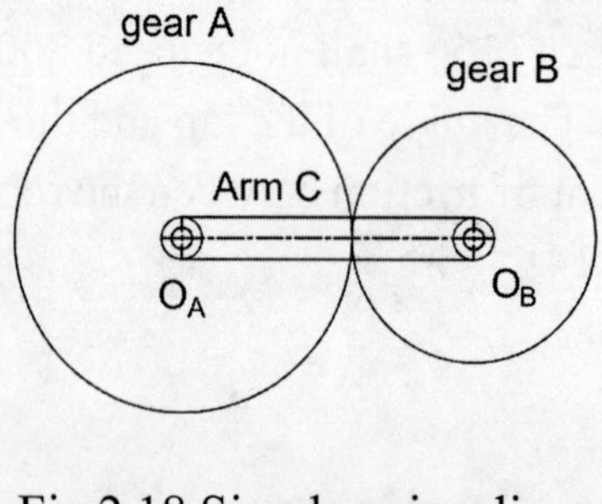

Fig.2.18 Simple epicyclic gear train.

Table 2.3 Motions of gear elements.

Step No	Conditions of motion	Revolutions of elements		
		Arm C	Gear A	Gear B
1	Arm C fixed. Gear A rotates through +1 revolution anticlockwise	0	+1	$-T_A/T_B$
2	Arm C fixed. Gear A rotates through + x revolutions	0	+ x	$-x(T_A/T_B)$
3	Add + y revolutions to all elements	y	y	y
4	Total Motion	y	x+y	$y-x(T_A/T_B)$

KNOW MORE

Watch videos on gear drives to understand the power transmission on https://www.youtube.com/playlist?list=PLuEePVBFb7EtegT-SGV9RgSBV5Po03CUt

Watch videos on Belt drives for the power transmission on https://www.youtube.com/watch?v=mQ14sL4v0Bs

Bibliography

- Theory of Machines, RS Khurmi and JK Gupta, S. Chand Publishing, 2005.
- Theory Of Machines, S. S. Rattan, McGraw Hill, 4th Edition, 2019.
- Theory of Mechanisms and Machines, Amitabha Ghosh and Asok Kumar Mallik, East-West Press Private Limited, 1998.
- Theory Mechanisms and Machine, Jagdish Lal, Metropolitan Book Pvt Ltd., 1994
- https://www.youtube.com/watch?v=mQ14sL4v0Bs

3 FLYWHEEL AND GOVERNORS

UNIT SPECIFICS

This unit describes the working principle and applications of flywheels and governors. Smoothening of variations due to cyclic operations by flywheel is explained. A few numerical examples help to understand the functioning of flywheel. The use of governors to maintain a constant speed due to variations in external loads is discussed. Comparison between flywheel and governor reviews their functions and their application to the reader.

RATIONALE

The flywheels and governors are used in controlling fluctuations of speeds due to cyclic load or external load respectively. The knowledge of flywheels and governors is essential for mechanical engineers in designing and manufacturing cams.

PRE-REQUISITE

Nil

UNIT OUTCOMES

The list of outcomes of this unit is as follows:

U3-O1: Understanding the concept and applications of flywheel.

U3-O2: Know about the governor used to minimise the fluctuation of speeds.

U3-O3: Comparison between flywheel and governor.

Unit Outcomes	**Expected Mapping with the Course Outcomes** (3- Weak Correlation; 2- Medium Correlation; 3- Strong Correlation)				
	CO-1	CO-2	CO-3	CO-4	CO-5
U3-O1	3	2	3	1	1
U3-O2	2	3	3	1	1
U3-O3	2	3	3	1	1

3.1 Concept of Flywheel

A flywheel is a mechanical rotating device that serves as a reservoir to store energy in a cycle of operation in machinery. The flywheel receives the excess energy during the cycle and releases it whenever the energy is demanded during operation. It has been observed that the flywheel speed increases as it absorbs the energy and decreases when it releases it. Therefore, a flywheel may not maintain a constant speed and simply fluctuate during a cycle or work.

In a few cases, like reciprocating steam engines and internal combustion engines, the energy is produced in one stroke and some energy is demanded in another stroke. In such cases, the flywheel smoothens the output torque. In mechanical applications like reciprocating compressors and pumps, more energy is required during the pressurization of fluid, and less energy is needed in the other part of the cycle. Machinery like riveting machines, punching machines, paper cutting machines in the press, etc., require a great amount of power during a short period. In such cases, the flywheel stores the energy and releases the required amount in one cycle of operation.

A flywheel controls the speed variations within a cycle operation. The fluctuation of the turning moment during each cycle due to processes is brought to a mean value. Therefore, the speed of a flywheel is not constant and varies around the mean speed. The flywheel carries a heavy rim so that enough kinetic energy is stored to maintain the mean torque, and the demand of energy to the system is released whenever it demands it. This flywheel is used for the smoothening of torque and speed of a prime mover or machinery.

3.2 Function and applications

A flywheel is a mechanical device attached to the machinery to control the speed variations caused by the fluctuation of the turning moment during each cycle of operation. The fluctuation may be caused by primary drivers or applications. Reciprocating steam engines, internal combustion engines, etc., are the prime movers while reciprocating compressors or pumps and punching machines are the applications.

In a four-stroke internal combustion engine, energy is produced during the power stroke, and no energy is developed during the other strokes, such as suction, compression, and exhaust. Energy has to be supplied to the engine to compress gasses. In such energy fluctuation cases, the flywheel is useful. It absorbs the excess energy during the power stroke by the flywheel and releases it during other strokes, thus rotating the crankshaft at a nearly uniform speed. The flywheel absorbs the extra energy available during function by increasing its speed. When it releases the energy for other purposes within the cycle of operation, the speed decreases. Therefore, a flywheel reduces the fluctuation of speed.

Working of the flywheel in 4-stroke engines

The turning moment diagram shows the variation of the turning moment of the engine with crank angle. Fig.3.1 shows the turning moment of a 4-stroke internal combustion engine with the crank angle for one cycle. There are four working strokes: suction, compression, power, and exhaust, completed in two revolutions of the crank, i.e. 4π radians. During the suction stroke, a small amount of energy is used to draw the air into the cylinder, resulting in a negative torque, and similarly, the next compression stroke also shows a negative torque that completes in one full rotation of the crank (2π) radians, as shown in the figure. The fuel burns and expands during the expansion or power stroke, resulting in a large positive torque or moment. The exhaust stroke drives out the burnt gases by utilising some energy from the system. That completes the one cycle of the engine.

The function of a flywheel fitted with this type of engine is to supply the energy for the suction of gasses, compress the gas, and drive out the gasses for working on an engine. The flywheel receives the energy from the engine during the expansion of gasses. The speed of the flywheel reduces while giving energy, resulting in a negative or less turning moment. Similarly, the speed of the flywheel increases while receiving the energy, resulting in a positive turning moment like in a power stroke of an engine.

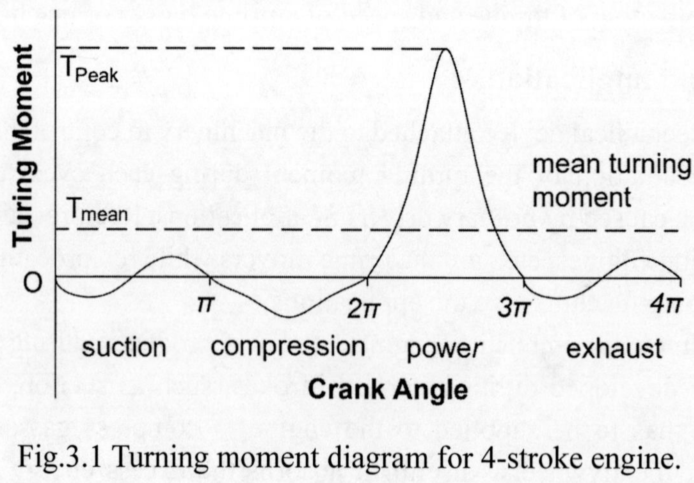

Fig.3.1 Turning moment diagram for 4-stroke engine.

3.3 Co-efficient of fluctuation of energy

A turning moment diagram for a multi-cylinder is shown in Fig.3.2. Let The horizontal line c-c represent the mean torque. These areas shown in the diagram represent some quantity of energy added or subtracted from the energy of the engine or flywheel. Let areas a1, a3, and a5 above the mean torque line show that the energy is added, and a2, a4, and a6 be the areas below the mean torque line, showing that the energy is received from the flywheel. The energy of the flywheel at different positions is as follows.

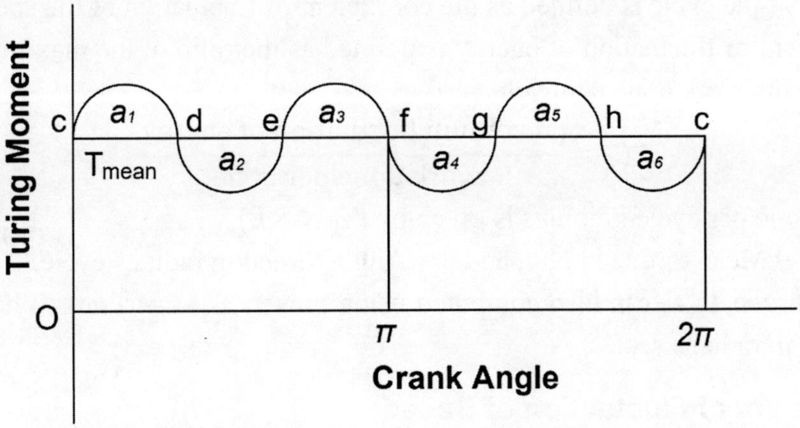

Fig.3.2 Turning moment diagram for a multi-cylinder.

Let E be the energy at the beginning point c

Energy at point c	= E
Energy at point d	= E + a1
Energy at point e	= E + a1 - a2
Energy at point f	= E + a1 - a2 + a3
Energy at point g	= E + a1 - a2 + a3 - a4
Energy at point h	= E + a1 - a2 + a3 - a4 + a5
Energy at point c	= E + a1 - a2 + a3 - a4 + a5 – a6
	= E (same as Energy at point c beginning)

Considering the two values of energies of the flywheel at point c,

$$a1 - a2 + a3 - a4 + a5 - a6 = 0$$

The greatest of these energies is the maximum kinetic energy of the flywheel, and the speed is the maximum corresponding to the crank position. In a similar way, the least of the

energies is the minimum kinetic energy of the flywheel, and the speed is the minimum corresponding to the other crank position.
The difference between the above maximum and minimum kinetic energies of the flywheel is called the maximum fluctuation of the energy. The ratio of the maximum fluctuation of energy to the work done in a cycle of operation is defined as the coefficient of fluctuation of the energy.
The difference between the maximum and minimum speeds of the flywheel is called as the maximum fluctuation of speed, and the ratio of the maximum fluctuation of energy to the mean speed of the cycle is defined as the coefficient of fluctuation of the speed.
The coefficient of fluctuation of energy is defined as the ratio of the maximum fluctuation of energy to the work done per cycle.

$$C_E = \frac{\text{Maximum fluctuation of energy}}{\text{work done per cycle}}$$

The Work done per cycle in joules is given by ($T_{mean} \times \theta$),
where T_{mean} = Mean torque in Nm, and θ = Angle turned in radians/cycle.
The mean torque T_{mean} can also computed using power. $T_{mean} = (P/\omega)$, Where Power P in watts and ω in radians/sec.

3.4 Coefficient of Fluctuation of Speed

The maximum fluctuation of speed is the difference between the maximum and minimum speeds during a cycle. The ratio of the maximum fluctuation of speed to its mean speed is called the coefficient of fluctuation of speed.

Let N_1 and N_2 be the maximum and minimum speeds in rpm during the cycle.
N be mean speed in rpm, i.e., $N = (N_1 + N_2)/2$
I is the mass moment of inertia of the flywheel about its centre
k is the radius of gyration of the flywheel

Coefficient of fluctuation of speed

$$C_s = \frac{N_1 - N_2}{N} = \frac{\omega_1 - \omega_2}{\omega} \qquad \text{Where } \omega \text{ in radians/sec.}$$

The coefficient of steadiness is reciprocal of C_s and is denoted by m

The mean kinetic energy of the flywheel $E = \frac{1}{2} I\omega^2$

Where mass moment of inertia $I = mk^2$

When the speed changes from ω_1 to ω_2, the change in fluctuation of energy is given by

$$\Delta E = \text{Maximum K.E.} - \text{Minimum K.E}$$

$$= \frac{1}{2} I\omega_1^2 - \frac{1}{2} I\omega_2^2 = \frac{1}{2} I(\omega_1^2 - \omega_2^2)$$

$$\Delta E = \frac{1}{2} I(\omega_1^2 - \omega_2^2)$$

$$= \frac{1}{2} I(\omega_1 - \omega_2)(\omega_1 + \omega_2)$$

By substituting $C_s = \frac{\omega_1 - \omega_2}{\omega}$ and $\omega = \frac{\omega_1 + \omega_2}{2}$

$$\Delta E = I\omega^2 C_s \qquad (1)$$

The size of the flywheel required for a service depends on the coefficient of fluctuation of speed and is an important limiting factor. The radius of gyration and mass of the rim of the flywheel are decided based on the value of the coefficient of fluctuation of speed.

Example 1: A flywheel is attached to an engine that has a mass of 6 tonnes, and the radius of gyration is 1.2 m. The maximum and minimum speeds in a cycle are observed to be 120 rpm and 115 rpm, respectively. Determine the maximum fluctuation of energy of the engine.
Solution.
Given: m = 6 T = 6000 kg, k = 1.2 m, N_1 = 120 rpm, N_2 = 115 rpm,

$$\text{Mean speed } N = \frac{(N_1 + N_2)}{2}$$

$$= \frac{(120+115)}{2} = 117.5 \text{ rpm}$$

$$\text{or} \quad \omega = \frac{2\pi N}{60} = \frac{2\pi(117.5)}{60} = 12.3 \text{ rad/s}$$

The coefficient of fluctuation of speed $C_S = \frac{2(N_1 - N_2)}{(N_1 + N_2)}$

$$= \frac{2(120-115)}{(120+115)} = 0.042$$

The fluctuation of energy $\Delta E = I\omega^2 C_s$

$$= (6000 \times 1.2^2)(12.3)^2(0.042)$$

$$= 54.9 \text{ kN-m}$$

Example 2 An engine develops a power of 200 kW at 180 rpm. The coefficient of energy fluctuation from the turning moment diagram is to be 0.1, and the fluctuation of speed is required to keep within ± 0.5% of the mean speed. The flywheel has a radius of gyration is 1.5 m. Find the mass of the flywheel required to meet the above requirements.
Solution.

Given: P – 200 kW; N – 180 rpm; C_E = 0.1; C_S = ± 0.5%; k = 1.5 m

The mean angular speed, $\omega = 2\pi N/60 = 2\pi \times 180/60 = 18.85$ rad/s

The fluctuation of speed is ± 0.5% of the mean speed; therefore, total fluctuation of speed,

$$(\omega_1 - \omega_2) = 1\%\ \omega = 0.01\ \omega$$

$$(\omega_1 - \omega_2)/\ \omega == 0.01 \quad \text{or} \quad C_S=0.01$$

Work done per cycle = 60P / N = 60 x 200 × 10^3 / 180 = 66.66 × 10^3 N-m

Maximum fluctuation of energy, ΔE = Work done per cycle × CE

= 66.66 × 10^3 × 0.1 = 6.66 × 10^3 N-m

the maximum fluctuation of energy $\Delta E = m.k^2.\omega^2.C_S$

$$6.66 \times 10^3 = m \times 1.5^2 \times (18.85)^2 \times 0.1$$

$$\therefore \quad m = 833.0 \text{ kg}$$

3.5 Governors

A governor is used to regulate the mean speed of internal combustion engines or prime movers like steam engines or turbines. The variations in the external load on the engine result in changes in the speed of the engines. If the engine's load increases, the engine's speed decreases, and it needs a correction to increase the speed so that the engine can deliver power at constant speed. This is achieved by increasing the supply of fuel to the engine. On the other hand, whenever the engine load comes less, the engine's speed increases. Thus, the governor senses the difference, and the working fluid is reduced in its action so that the engine speed is normal.

The governor controls the supply of fuel to the engine according to the varying load conditions and keeps the mean speed within specific limits. A mechanism is used to operate the fuel valve, specifically when the load increases or decreases. The governor's configuration changes with the engine's speed, and drives the fuel valve to increase the fuel supply. Similarly, when the load decreases, the engine speed increases and the governor initiates a decrease in the supply of fuel to the engine.

We know that the flywheel minimises the fluctuations in engine speed within the cycle. It cannot alter the fluctuations due to load variation. Therefore, a flywheel does not exercise any control over the mean speed of the engine in a cycle. A governor is used to minimise the fluctuations in the mean speed due to the external load variation. The governor has no influence over cyclic speed fluctuations. However, it controls the mean speed over a long period during which the load on the engine may vary. Therefore, a governor automatically regulates the fuel supply to the engine as demanded by variation of load, through linkages so that the engine speed is maintained nearly constant

3.6 Types of Governors

The governors are classified broadly into Centrifugal governors and Inertia governors. The centrifugal governors may be classified further as pendulum type and loaded type. The centrifugal governors work on the balancing of centrifugal force on the rotating balls by an equal and opposite radial force. The Watt governor is an example of a pendulum type. Meanwhile, deadweight governors and spring-loaded governors are the centrifugal governors. The Porter and Proell governors are the deadweight type of governors. Hartnell's governor is a spring-loaded type of governor. Fig.3.3. shows these governors: Centrifugal type Governor: Watt Governor, Gravity Loaded Governor: Porter Governor and Proell Governor and Spring-loaded Governor: Hartnell Governor.

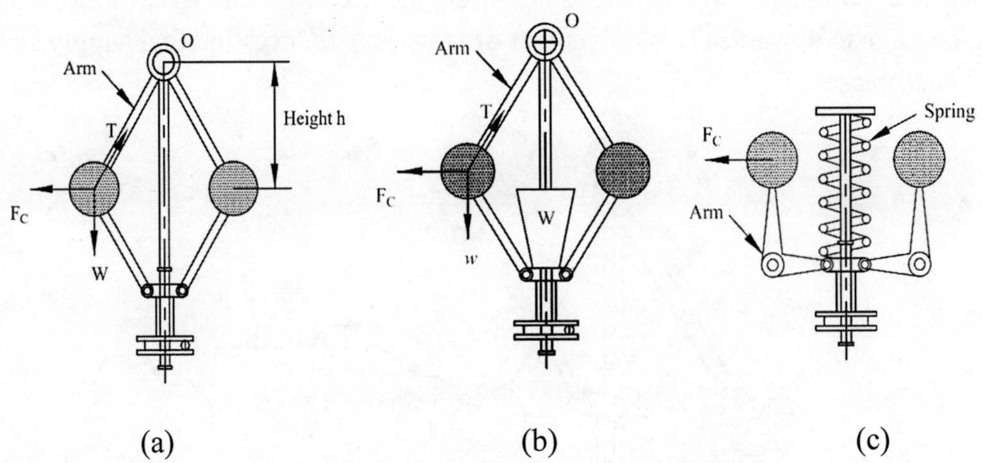

Fig.3.3. Type of Governors (a) Watt Governor, (b) Gravity Loaded Governor, and (c) Spring-loaded Governor: Hartnell Governor

3.7 Centrifugal Governor

The centrifugal governors work on the principle of balancing of centrifugal force on the rotating balls by an equal and opposite radial force. The governor consists of two balls of equal mass, hinged with links and arms, as shown in Fig. 3.3(a).

These balls are known as flyballs, which revolve with a spindle. The spindle of the governor in driven by the engine. The ends of the arms are pivoted to the spindle at the top and the other end to flyballs. While revolving about the vertical axis, the balls may rise up or fall down as they. The balls are connected with the links, and the other end is to the sleeve. This sleeve also revolves around the spindle and can slide up and down.

When the spindle speed increases, the flyballs move away from the spindle axis and lift the sleeve. Similarly, the sleeve comes down or falls when the speed decreases. The upward and downward travel of the sleeve is limited by the two stoppers provided on the spindle. This corresponds to the maximum and minimum speed of the engine. The sleeve drives the throttle valve by using a bell crank lever, as shown in the figure. The supply of the fuel reduces when the sleeve rises and increases when it falls.

The engine and governor's speed decreases when the engine's load increases. Thus, the centrifugal force on flyballs reduces, resulting in the flyballs moving inwards and pushing the sleeve downward. The downward movement of the sleeve opens the throttle valve to increase the supply of fuel, and thus, the engine speed is increased till it balances with normal speed. Similarly, when the load on the engine decreases, the flyballs move outwards and the sleeve to upwards. This movement of the sleeve reduces the fuel supply so that the speed is decreased.

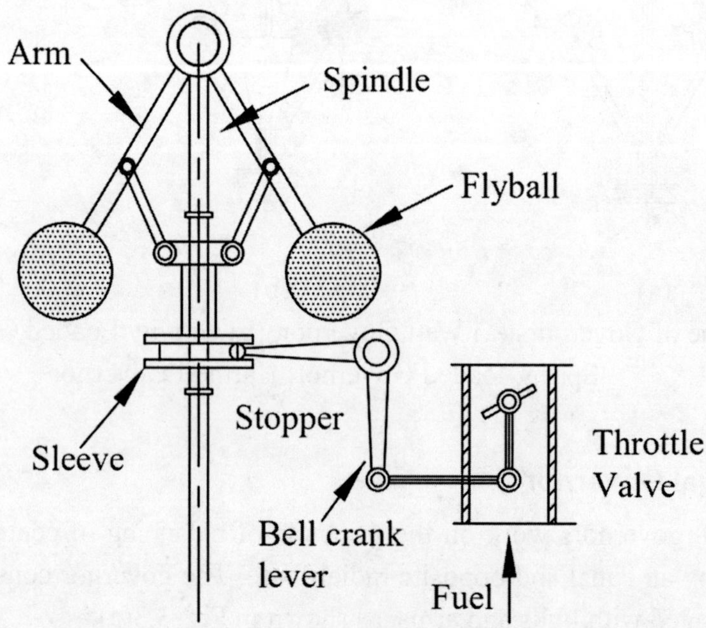

Fig.3.3(a) Centrifugal Governor.

Terms used in governors:

The following are the terms used generally used in connection with the governors.

Height of a governor: A vertical distance from the centre of the ball to a point where the axes of the arms intersect on the spindle axis

Equilibrium speed: The speed at which the governor balls, arms etc., are in complete equilibrium

Mean equilibrium speed: The speed at the mean position of the balls or the sleeve.

Maximum and minimum equilibrium speeds: The speeds at the maximum and minimum radius of rotation of the balls respectively.

Sleeve lift: The vertical distance travelled by sleeve travels due to change in equilibrium speed.

Centrifugal force: The force developed due to the rotation of flyballs.

Governor Height: The vertical height of fly balls from the arm pivot

Sensitiveness of the governor is defined as the ratio of the difference between the maximum and minimum equilibrium speeds to the mean equilibrium speed.

Stability of a governor is said to be stable when there is one radius of rotation of the balls for each speed which is within the speed range of the governor.

Isochronism of a governor is said to be isochronous if the equilibrium speed is constant for all the radii of rotation of balls within the working range.

Hunting of a governor is that the speed of an engine fluctuates due to the change in load on the engine.

3.8 Watt Governor

A Watt governor is the simplest form of a centrifugal governor and is shown in Fig.3.4. It is basically a pendulum with links of negligible mass pivoted at the top and connected to a sleeve at the bottom. Let the arm be pivoted at the spindle as at point O, as shown (in case the pivot point is offset, the intersecting point on the axis be considered). If the speed increases, the flyballs move radially outward till they reach the position of equilibrium and continue in the same status in the equilibrium. Consider the forces acting on one-half of the governor as shown in the figure.

Let m = Mass of the ball in kg,

w = Weight of the ball in N

T = Tension in the arm in N

ω = Angular velocity of spindle axis in rad/s,

r = Radius of the path of rotation of the ball in m

h = Height of the governor in m

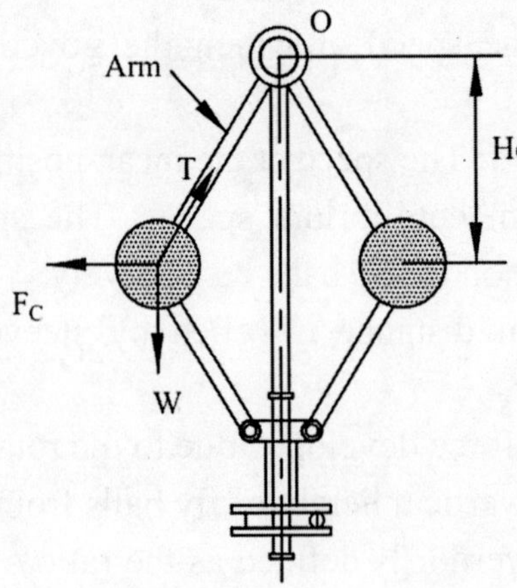

Fig.3.4 Watt Governor.

Consider the situation where the ball is in equilibrium under the action of

- The centrifugal force ($F_C = m\omega^2 r$) acting on the ball radially outward,
- The tension (T) in the arm, and
- The weight (w) of the ball.

Taking moment about point O, we have

$$F_C \times h = w \times r = (mg)\, r$$

or $$(m\omega^2 r)\, h = (mg) r$$

$$\mathbf{h = g/\omega^2} \qquad (2)$$

The height of a governor h is inversely proportional to (speed)^2. Therefore, the value of 'h' is small as speed increases. The change in the value of h corresponding to the speed change gives the necessary change in the fuel supply. This type governor may work satisfactorily for low-speed ranges.

Example 3: A Watt governor fitted to an engine is rotating at 80 rpm and is in equilibrium. If the speed of engine has changed to 78 rpm, Calculate the vertical height of the sleeve.

Solution:

Given: $N_1 = 80$ rpm, $N_2 = 78$ rpm.

$$\omega_1 = 2\pi N_1/60 = 2\pi \times 80/60 = 8.378 \text{ rad/s}$$
$$\omega_2 = 2\pi N_2/60 = 2\pi \times 78/60 = 8.169 \text{ rad/s}$$

Change in height $h = h_1 - h_2$

or $$h = \frac{g}{\omega 1^2} - \frac{g}{\omega 2^2}$$

$$= \frac{9.81}{8.378^2} - \frac{9.81}{8.169^2} = -7.24 \times 10^{-3} \text{ m}$$

3.9 Porter Governor

The Watt governor is modified, adding a central weight that slides along with the sleeve. The Porter Governor is shown in Fig.3.5. The central weight moves up and down according to the speed variation along the central spindle. The additional downward force due to central weight increases the speed of revolution, which enables the flyballs to rise to the predetermined level. The forces acting on one-half of the governor are shown below.

Let m = Mass of each ball in kg,

M = Mass of the central load in kg,

r = Radius of rotation in m,

h = Height of governor in m,

N = Speed of the balls in rpm,

ω = Angular speed of the balls in rad/s,

F_C = Centrifugal force acting on the flyballs in N

T_1 and T_2 = Force in the arm and link in N,

α and β = Angle of inclination of the arm and link with the vertical,

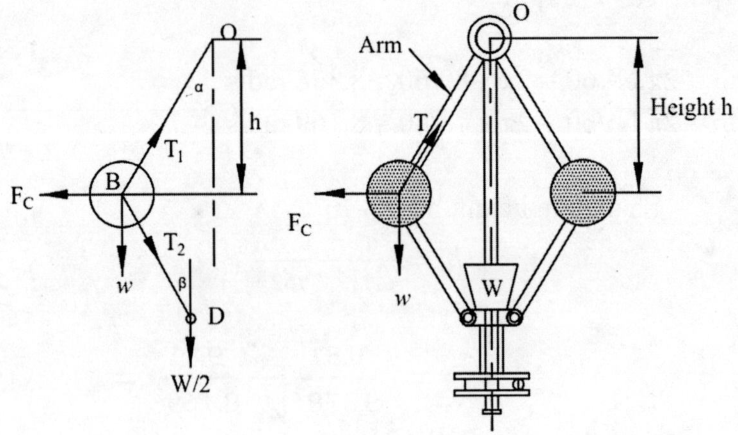

Fig.3.5 Porter Governor.

At point D, two forces are acting Tension in link T_2 and weight of mass M that is (W = Mg). Assuming that the governor is in equilibrium with the speed of governor N_1 and resolving forces vertically, we get

$$T_2 \cos\beta = Mg/2$$

$$T_2 = \frac{Mg}{2\cos\beta} \qquad (3)$$

Again, consider the equilibrium of the forces acting on point B. The following are the forces act on this point B and are as shown in figure above.

The weight of the ball (w = mg), the Centrifugal force F_C that is ($F_C = m\omega^2 r$), The tension in the arm T_1, and the link T_2.

Resolving all above the forces vertically, and substitute T_2 from eqn (3)

$$T_1 \cos\alpha = T_2 \cos\beta + w = \frac{Mg}{2} + mg \qquad (4)$$

Resolving all above the forces horizontally,

$$T_1 \sin\alpha + T_2 \sin\beta = F_c$$

$$T_1 \sin\alpha + \frac{Mg}{2\cos\beta}\sin\beta = F_c$$

$$T_1 \sin\alpha + \frac{Mg}{2}\tan\beta = F_c$$

or $$T_1 \sin\alpha = F_c - \frac{Mg}{2}\tan\beta \qquad (5)$$

Divide above Eqn. (2) by (3)

$$\frac{T_1 \sin\alpha}{T_1 cos\alpha} = \frac{F_c - \frac{Mg}{2}\tan\beta}{\frac{Mg}{2} + mg}$$

or
$$\tan\alpha\left[\frac{Mg}{2} + mg\right] = F_c - \frac{Mg}{2}\tan\beta$$

$$\left[\frac{Mg}{2} + mg\right] = \frac{F_c}{\tan\alpha} - \frac{\text{Mg}\tan\beta}{2\tan\alpha}$$

Now substitute $q = \frac{\tan\beta}{\tan\alpha}$ and tan α = r/h, we get

$$\left[\frac{Mg}{2} + mg\right] = m\omega^2 r\,\frac{h}{r} - \frac{\text{Mg}}{2}q$$

$$m\omega^2\text{h} = \frac{Mg}{2}(1 + \text{q}) + \text{mg}$$

Therefore

$$\text{h} = \left[\frac{Mg}{2}(1 + \text{q}) + \text{mg}\right]\frac{1}{m\omega^2}$$

$$\text{h} = \frac{m + \frac{M}{2}(1+\text{q})}{m} \text{x} \frac{\text{g}}{\omega^2} \qquad (6)$$

$$\left(\frac{2\pi N}{60}\right)^2 = \frac{m + \frac{M}{2}(1 + \text{q})}{m} \text{x} \frac{9.81}{h}$$

$$(N)^2 = \frac{m + \frac{M}{2}(1+\text{q})}{m} \text{x} \frac{895}{h} \qquad (7)$$

if the arm and link are equal, the q=1. Then, $N^2 = \frac{m+M}{m} \text{x} \frac{895}{h}$

Note: 1) The sensitiveness of the governor is defined as the ratio of the difference between the maximum and minimum equilibrium speeds to the mean equilibrium speed. Let N_1 and N_2 be the maximum and minimum speed of the engine to which the governor is attached. Then mean speed is $(N_1 + N_2)/2$.

∴ Sensitiveness of the governor = $\frac{2(N_1 - N_2)}{(N_1 + N_2)}$ (8)

2) The stability of a governor is said to be stable when there is one radius of rotation of the balls for each speed, which is within the speed range of the governor.
3) The Isochronism of a governor is said to be isochronous if the equilibrium speed is constant for all the radii of the rotation of balls within the working range. Therefore, the speed range is zero for an isochronous governor, and this type of governor shall maintain constant speed. The isochronism is the stage of infinite sensitivity.
4) Hunting of a governor: The speed of an engine changes due to the change in load on the engine. Then, the sleeve has a tendency to move to a new position, so it overshoots the desired position. Sleeve then moves back but again overshoots the desired position due to inertia. This results in the setting up of oscillations in engine speed. The governor, then, tends to speed variation instead of controlling it. This phenomenon is known as the hunting of the governor.

Example 4: A Porter governor is used to control the speed of an engine. It has an equal arm and is 200 mm long. Links are pivoted on the axis of rotation. Two balls of mass of 5 kg each and a central load of mass 20 kg on the sleeve are used. The radius of rotation of the balls is 125 and 140 mm when the governor begins to lift at minimum and maximum speed. Find the minimum and maximum speeds and range of speed of the governor.
Solution:
Given: L=200 mm, q = 1, r_1 = 125 mm, r_2 = 140 mm, m = 5 kg, M = 20 kg
Sketching the governor configuration for two speeds,

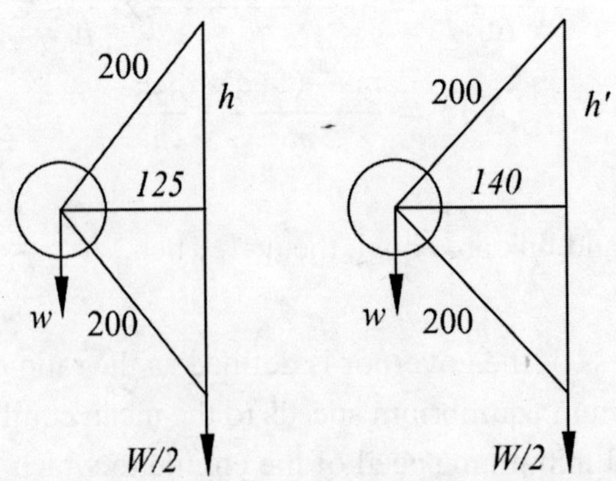

Fig.3.6 Governor configuration for Example 4.

Height of balls at minimum and maximum speeds

$$h = \sqrt{200^2 - 125^2} = 156.mm = 0.156 \text{ m}$$

and

$$h' = \sqrt{200^2 - 140^2} = 142.mm = 0.142\ m$$

For Porter Governor, we use relation

$$(N)^2 = \frac{m+M}{m} \text{X} \frac{895}{h}$$

Substituting values, we get $(N)^2 = \frac{5+20}{5} \text{X} \frac{895}{0.152}$ or N=169.3 rpm

Similarly, for other speeds of governor

$$(N')^2 = \frac{5+20}{5} \text{X} \frac{895}{0.142} \quad \text{or N'= 177 rpm}$$

Range of speed of governor 177-169.3 = 7.7 rpm

Example 5: A Porter governor is used to control the speed of an engine. It has an equal arm and is 250 mm long. Links are pivoted on the axis of rotation. Two balls of mass of 5 kg each and a central load of mass 20 kg on the sleeve are used. The radius of rotation of the balls is 125 and 140 mm when the governor begins to lift at minimum and maximum speed.

a) Find the minimum and maximum speeds and range of speed of the governor.
b) Determine the speed range if the friction at the sleeve is equivalent to 10 N.

Given: L = 250 mm, r_1 = 100 mm, r_2 = 120 mm, m = 2 kg, M = 15 kg, F_W = 10 N

Sketching the governor configuration for two speeds,

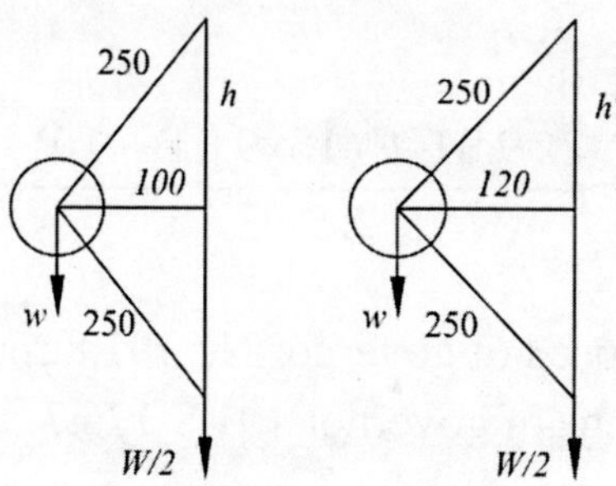

Fig.3.7 Example 5.

(a) Porter governor without sleeve friction

Height of balls at minimum and maximum speeds

$$h = \sqrt{250^2 - 100^2} = 156.mm = 0.229 \text{ m}$$

and

$$h' = \sqrt{250^2 - 120^2} = 142.mm = 0.219\ m$$

For Porter Governor, we use relation

$$(N)^2 = \frac{m+M}{m} \text{x} \frac{895}{h}$$

Substituting values, we get $(N)^2 = \frac{2+15}{2} \text{x} \frac{895}{0.229}$ or N=182.2 rpm

Similarly for other speed of governor

$$(N')^2 = \frac{5+20}{5} \text{x} \frac{895}{0.219} \quad \text{or N'= 186.2 rpm}$$

Speed Range of governor 186.2-182.2 = 4 rpm

(b) Porter governor with sleeve friction

When sleeve move downward, then

$$(N)^2 = \frac{mg + (Mg - fw)}{mg} \text{x} \frac{895}{h}$$

Substituting the values, we get

$$(N)^2 = \frac{2\text{x}\,9.81 + (15\text{x}9.81 - 10)}{2\text{x}9.81} \text{x} \frac{895}{0.229}$$

Speed of governor N= 176.7 rpm

When the sleeve move upward, then

$$(N)^2 = \frac{2\text{x}\,9.81 + (15\text{x}9.81 + 10)}{2\text{x}9.81} \text{x} \frac{895}{0.219}$$

Speed of governor N= 191.8 rpm

Speed Range of governor 191.8-176.7 = 15.1 rpm.

Unit Summary

- A flywheel is a mechanical rotating device that serves as a reservoir to store energy in machinery. It has been observed that the flywheel speed increases as it absorbs the energy and decreases when it releases it.
- The flywheel receives the excess energy during the cycle and releases it whenever the energy is demanded during operation. The flywheel carries a heavy rim so that enough kinetic energy is stored to maintain the mean torque. This flywheel is used for the smoothening of torque and speed of a prime mover or machinery.
- Application of flywheels are found in the prime movers like reciprocating steam engines or internal combustion engines to smoothen the working cycle and in reciprocating compressors or pumps, and punching machines to store the energy and release it in a short time during working.
- The size of the flywheel required is depend on the coefficient of fluctuation of speed and the radius of gyration.
- A governor is used to regulate the mean speed of engines or turbines. The variations in the external load on the engine result in changes in the speed of the engines. The speed of the engine is driven to its normal speed using a governor.
- Governors use flyballs and the position of these ball depend on the speed. The lift of the sleeve is used to operate throttle or fuel pump. If speed of the engine decrease, the governor senses it and correspondingly increase the fuel, so that the engine speed has come to normal.
- The governors are classified broadly into Centrifugal governors and Inertia governors. The centrifugal governors work on the balancing of centrifugal force on the rotating balls by an equal and opposite radial force. The Watt governor is an example of a pendulum type.
- The centrifugal governors work on the principle of balancing of centrifugal force on the rotating balls by an equal and opposite radial force.
- The height of a Watt governor h is inversely proportional to $(\text{speed})^2$. Therefore, the value of 'h' is small as speed increases.
- The Watt governor is modified to a Porter Governor, adding a central weight that slides along with the sleeve. The central weight moves up and down according to the speed variation along the central spindle. The additional downward force due to central weight increases the speed of revolution.

Multiple Choice Questions

1. A flywheel is sensitive to the fluctuations of
   (a) load on the engine (b) cyclic variation of the engine (c) Speed of the engine (d) fuel in the engine

2. The kinetic energy of a flywheel is
   (a) Iw^2 (b) Iw (c) $Iw^2/2$ (d) $Iw/2$

3. The speed of a flywheel when a driving torque is more than load torque,
   a) Accelerated, b) Decelerated c) Constant d) Zero

4. The coefficient of fluctuation of the speed of a flywheel is given by
   (a) $\frac{\omega_1-\omega_2}{\omega_1+\omega_2}$ (b) $\frac{\omega_1+\omega_2}{\omega_1-\omega_2}$ (c) $\frac{2(\omega_1-\omega_2)}{\omega_1+\omega_2}$ (d) $\frac{\omega_1-\omega_2}{2(\omega_1+\omega_2)}$

5. A flywheel rotating at 5 rad/s has a mass of 50 kg with a radiation of gyration of 0.15 m. The rotational kinetic energy of the flywheel is
   (a) 37.5 (b) 18.75 (c) 2.81 (d) 5.62

6. The function of a Governor is to respond to the fluctuations of
   (b) speed of an engine (b) operating cycle of the engine (c) internal pressure of cylinder of the engine (d) fuel level in the engine

7. The sensitiveness of a governor is given by
   (a) $\frac{\omega_1-\omega_2}{\omega_{mean}}$ (b) $\frac{\omega_1+\omega_2}{\omega_{mean}}$ (c) $\frac{\omega_1-\omega_2}{2\,\omega_{mean}}$ (d) $\frac{2(\omega_1-\omega_2)}{\omega_{mean}}$

8. The height of the governor is in a centrifugal governor
   (a) decreases with speed (b) increases with speed (c) do not vary with speed (d) fixed for all speeds.

9. The Sleeve of a Porter governor moves upwards when the speed of the engine is
   (a) increases, (b) decreases (c) constant (d) zero

10. The Porter governor is
(a) spring loaded governor (b) inertia governor (c) centrifugal governor (d) flywheel governor

Answers to Multiple Choice Questions

1. (b), 2. (c), 3. (a), 4. (c), 5. (c), 6. (a), 7. (a), 8. (b), 9. (a), 10. (c)

Exercises

1. A steam engine running at 120 rpm develops a power of 125 kW. The from the turning moment diagram for the engine shows that the coefficient of energy fluctuation to be 0.15, and the fluctuation of speed is required to keep within ± 1.0% of the mean speed. The flywheel has a radius of gyration is 1.10 m. Find the mass of the flywheel required to meet the above requirements. [2453 kg].

2. A flywheel is attached to an engine has a mass of 5 tonnes and the radius of gyration is 0.75 m. The maximum and minimum speeds in a cycle is observed to be 120 rpm and 110 rpm respectively. Determine the maximum fluctuation of energy of the engine. [35.48 kN-m].

3. A punching machine has a flywheel and is running at mean speed of 20 m/s. It is used to produce holes of size 40 mm in a sheet of 35 mm and requires 8 Nm of energy per sq. mm of shear area. The stroke of tool in punching machine is 95 mm. The total fluctuation of speed is limited to 3%. Determine the mass of the flywheel. [2392 kg].

4. A steam engine develops 100 kW at a rated speed of 80 rpm and is fitted with a flywheel of radius of gyration is 0.75 m. The maximum fluctuation of energy Cs is 0.30 of the work done per stroke. The maximum and minimum speeds are limited to one per cent on either side of the mean speed. Find the mass of the flywheel required, [2848 kg].

5. A Watt governor fitted to an engine is rotating at 70 rpm and is in equilibrium. If the speed of engine has changed to 68 rpm, Calculate the vertical height of the sleeve due to change in engine speed [0.01m].

6. A Watt governor fitted to an engine is rotating at 120 rpm and is in equilibrium. If the speed of engine has changed to 135 rpm, Calculate the vertical height of the sleeve due to change in engine speed [0.013m].

7. A Porter governor has an equal arm and is 180 mm long. Links are pivoted on the axis of rotation. Two balls of mass of 4 kg each and a central load of mass 16 kg on the sleeve are used. The radius of rotation of the balls is 130 and 145 mm when the governor begins to lift at minimum and maximum speed. Find the minimum and maximum speeds and range of speed of the governor. [15.25 rpm].

8. A Porter governor is used to control the speed of an engine. It has an equal arm and is 275 mm long. Links are pivoted on the axis of rotation. Two balls of mass of 6 kg each and a central load of mass 25 kg on the sleeve are used. The radius of rotation of the balls is 150 and 160 mm when the governor begins to lift at minimum and maximum speed. Find the range of speed of the governor. [2.31 rpm].

9. A Porter governor has a mass of the central load is 18 kg and the mass of each ball is 2 kg. The top and bottom arms are 250 and 300 mm respectively. The friction of the sleeve during travel is 14 N. The top arms make 45° with the axis of rotation in the equilibrium position. Find the range of speed of the governor in that position [15 rpm].

10. A Porter governor with upper and lower arms each 250 mm long are pivoted on the axis of rotation. The mass of each rotating ball is 3 kg and the mass of the sleeve is 20 kg. The sleeve is in its lowest position when the arms are inclined at 30° to the governor axis. The lift of the sleeve is 36 mm. Find the range of speed of the governor. [16 rpm].

Experiment

Aim: To study the effect of ball mass and weight of the centre of sleeve in Porter Governor

Procedure:

- Note down the effective lengths of links, weights of flying balls, and central weight.
- Assemble all the essential links and weights according to the type of governor.
- Tighten the necessary bolts and confirm.
- Start the motor and gradually increase the speed.
- The flyballs will move outward due to which the sleeve will also rise.
- Note down the speed and sleeve rise
- Repeat the experiment at different speeds till the balls fly to maximum position.
- Bring back the sleeve down by reducing the speed gradually and stop

Specifications:

Mass of Governor Balls (m) in kg

Mass of central weight (M) in kg

Length of the link L_1 (m)

Length of the link L_2 (m)

Initial height h_o (m)

Ratio of links (L_1/L_2)

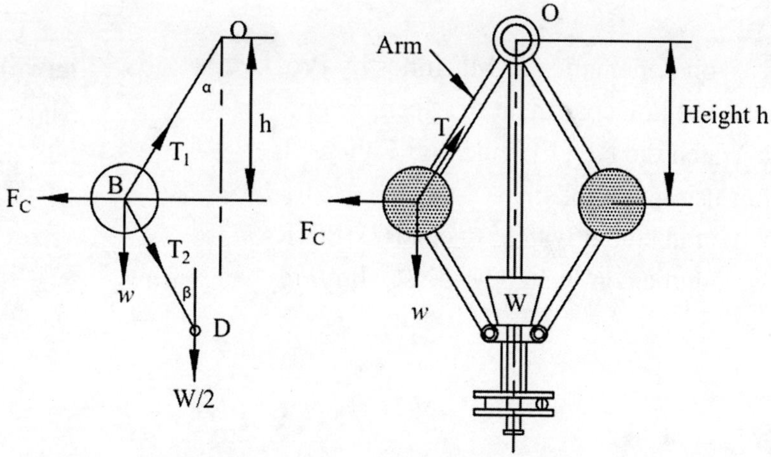

Fig.3.8 Porter governor.

Table 3.1 Observations on Porter Governor Experiment.

Sl.No	Speed (rpm)	Speed (rad/s)	Sleeve lift (mm) (experimental)	Sleeve lift (mm) (theoretical)

For theoretical calculations, use the equation for governor height

$$h = \frac{m + \frac{M}{2}(1 + q)}{m} x \frac{g}{\omega^2}$$

Plot a graph height vs speed to know the sensitivity of the Governor and the corresponding speed.

KNOW MORE

Lecture Series on Kinematics of Machines by Prof. Asok Kumar Mallik, Department of Mech. Engg. IIT Kanpur. Watch the NPTEL video on Flywheels and Governor using the links:

https://www.youtube.com/watch?v=oZhR1HPdvR4

https://www.youtube.com/watch?v=dXsQslmyxf4

Bibliography

- Theory of Machines, RS Khurmi and JK Gupta, S. Chand Publishing, 2005.
- Theory Of Machines, S. S. Rattan, McGraw Hill, 4th Edition, 2019.
- Theory of Machines and Mechanisms, John J. Uicker et al., Oxford University Press, Fifth Edition, 2017.
- Theory of Mechanisms and Machines, Amitabha Ghosh and Asok Kumar Mallik, East-West Press Private Limited, 1998.
- Theory of Mechanisms & Machines [Paperback] Dr ...
- Theory Mechanisms and Machine, Jagdish Lal, Metropolitan Book Pvt Ltd., 1994
- https://www.youtube.com/watch?v=oZhR1HPdvR4
- https://www.youtube.com/watch?v=dXsQslmyxf4

4 BRAKES, DYNAMOMETER, CLUTCHES AND BEARINGS

UNIT SPECIFICS

This unit presents the function and applications of brakes and dynamometers in the first part. Construction and working of various brakes and the concept of self-locking & self-energizing brakes are included. Numerical problems to find the braking force and torque help visualise the working of brakes and dynamometers. Uniform pressure and Uniform Wear theories and the function of clutches are focused on their applications. The use of bearings in most of the equipment and topics is useful for engineers. Therefore, this unit is very important for readers from the engineering point of view.

RATIONALE

The use of elements like brakes and dynamometers, as well as clutches and bearings, are common in every machinery. The knowledge of these are essential for mechanical engineers in designing and manufacturing.

PRE-REQUISITE

Nil

UNIT OUTCOMES

The list of outcomes of this unit is as follows:

U4-O1: Understand the concept and application of brakes and dynamometers.

U4-O2: Concept of Uniform Pressure and Uniform Wear theories for clutches.

U4-O3: Understand the concept and application of bearings.

Unit Outcomes	**Expected Mapping with the Course Outcomes** (4- Weak Correlation; 2- Medium Correlation; 3- Strong Correlation)				
	CO-1	CO-2	CO-3	CO-4	CO-5
U4-O1	3	3	3	1	1
U4-O2	3	3	3	1	1
U4-O3	3	3	3	1	1

4.1 Brakes and dynamometers

A mechanical brake is a device used to retard or stop the motion of a machine member purposefully. Brakes are used extensively in automobiles, brakes in hoists, elevators, etc., or to gradually slow down the motion of members. A moving body has a contact surface on which it slides, and a coat of an artificial frictional layer is applied. When a force is applied to this moving machine member, its motion retards and stops relative to the other member. A frictional force is developed at this contact surface, resulting in a heated surface. This heat is due to the conversion of either the kinetic energy of the moving member or potential energy. The effectiveness of a brake system depends on the unit pressure applied between the braking surfaces and the coefficient of friction between the braking surfaces. While, a dynamometer is a type of brake, it has an additional device to measure the frictional resistance so that the device can be used to assert the torque transmitted and, hence, the power.

4.2 Types of brakes and dynamometers

Brakes are classified according to the energy transformation in brake elements;

a) Hydraulic brakes, b) Electric brakes, and c) Mechanical brakes.

The mechanical brakes are classified according to the direction of the acting force

(a) Radial brakes and axial brakes
(b) block or shoe brakes and band brakes
(c) External brakes and internal brakes.
(d) Disc brakes and cone brakes.

There are two types of Dynamometers

1. Absorption dynamometers,
2. Transmission dynamometers.

4.3 Comparison between brakes and dynamometers

Brakes and dynamometers are different mechanical devices. A brake is a mechanical device that constrains motion by absorbing energy from a moving system. On the other hand, a dynamometer is used to measure the tangential force or mechanical power. It is used to measure the output of power of a rotating member.

Table 4.1 Comparison between brakes and dynamometers

Brakes	**Dynamometers**
Brakes work on converting kinetic energy into heat, which is dissipated to the surroundings.	Dynamometers work on the principle that power is measured and converted into heat by frictional resistance.
It offers frictional resistance to the moving body to bring the moving body to the rest.	It can measure the frictional resistance, which is the measure of power.
It is used to retard or stop the vehicle or a moving body.	It is used to measure the tangential force on a rotating body.
Examples: Block and band brakes, internal expanding shoe brakes, and Hydraulic brake systems.	Examples: Rope brake dynamometers, hydraulic dynamometers, Eddy current dynamometers

4.4 Shoe Brake

A single block or shoe brake consisting of a block pressed against the rim of a revolving brake wheel drum is shown in Fig. 4.1(a). Usually, the brake blocks are made with softer material than the rim of the wheel. When a force P is applied at the end of the lever, it presses the wheel, causing a frictional resistance at the area of contact subtended by an angle 2θ. The frictional force between the block and the wheel causes a tangential force called braking force to act on the wheel tangentially. This results in retard the rotation of the wheel. It works efficiently with small wheels, and an angle of contact is less than 60o and is commonly used in railway trains, tram cars, trolleys on guideways, etc.

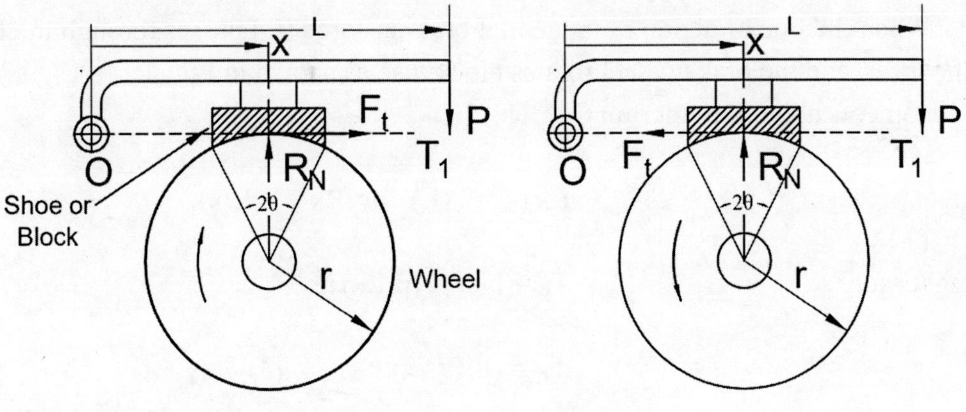

Fig.4.1(a) Shoe brake.

Let P = Force applied at the end of the lever,
R_N = Normal force pressing the brake block on the wheel,
F_t = Tangential braking force
r = Radius of the wheel,
2θ = Angle of the contact surface,
μ = Coefficient of friction,
L = distance of force P from point O
x = distance of force R_N from point O

The braking torque, acting at the contact surface of the shoe and the wheel, is

$$T_B = F_t.r \qquad (1)$$

Assuming the normal pressure between the block and the wheel is uniform, then, the tangential braking force on the wheel,

$$F_t = \mu.R_N$$

Therefore, the braking torque,

$$T_B = F_t.r = (\mu.R_N).r \qquad (2)$$

Let us now consider the three cases that possibly depend on the location of the fulcrum of the lever: The line of action of tangential force F_t passes through point O, or above or below the fulcrum point O. The configuration of these three cases is shown on figures, 4.1a) to c).

Case1: When the line of action of tangential braking force (F_t) passes through the fulcrum O of the lever, and the brake wheel rotates clockwise as shown in Fig. 4.1 (a),
Taking moments about the fulcrum O,

we get $R_N(x) = P(L)$ or $R_N = P(L/x)$

Braking torque, $T_B = F_t.r = (\mu.R_N).r$

$$\mathbf{T_B = (\mu.PL/x).r} \qquad (3)$$

If the brake wheel rotates anticlockwise, as shown in Fig.4.1(a), the braking torque is the same as above. The direction of F_T is opposite if the wheel rotates in a clockwise direction.

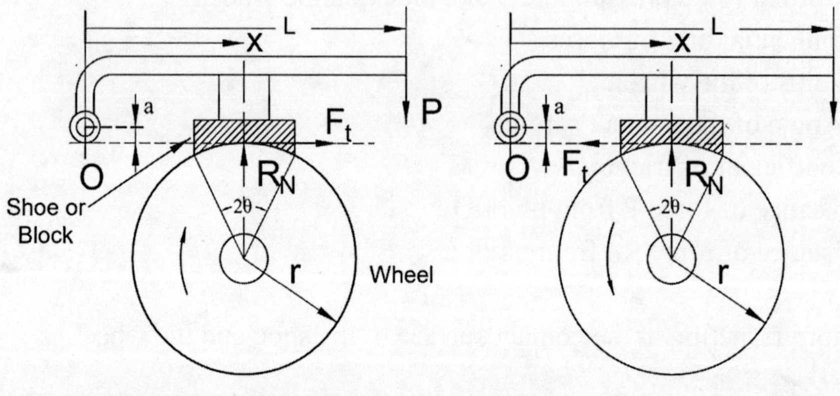

Fig.4.1(b) Line of action of friction is above the point O.

Case 2: When the line of action of tangential braking force (F_t) is at a distance "a" from the fulcrum O of the lever, Refer to the brake wheel rotates clockwise and also counterclockwise, as shown in Fig. 4.1(b),

Taking a moment about point O,

Then, $R_N(x) + Ft(a) = P(L)$

or $R_N(x) + \mu.R_N(a) = P(L)$

$$R_N(x + \mu.a) = P(L)$$

$$\text{Therefore,} \quad T_B = \frac{\mu PLr}{(x + \mu a)} \qquad (4)$$

When the brake wheel rotates anticlockwise,

Taking a moment about point O,

$$\text{Then,} \quad R_N(x) = Ft(a) + P(L)$$

$$\text{or} \quad R_N(x) = \mu.R_N(a) + P(L)$$

$$R_N(x - \mu.a) = P(L)$$

$$\text{Therefore,} \quad T_B = \frac{\mu PLr}{(x - \mu a)} \qquad (5)$$

Case 3: When the line of action of tangential braking force (F_t) is at a distance "a" above the fulcrum O of the lever, Refer to the brake wheel rotates clockwise and also counterclockwise, as shown in Fig. 4.1(c),

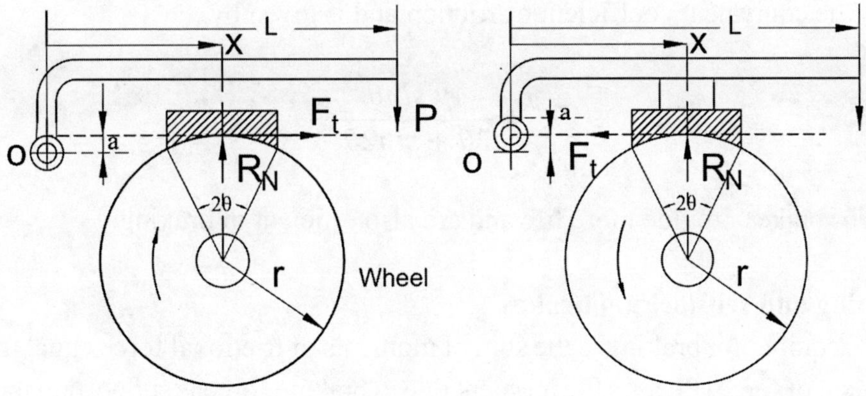

Fig.4.1(c) The Line of action of friction is above the point O.

Taking a moment about point O,

$$\text{Then,} \quad R_N(x) = Ft(a) + P(L)$$

$$\text{or} \quad R_N(x) = \mu.R_N(a) + P(L)$$

$$R_N(x - \mu.a) = P(L)$$

$$\text{Therefore,} \quad T_B = \frac{\mu PLr}{(x - \mu a)} \qquad (6)$$

When the brake wheel rotates anticlockwise, as shown in Fig. 4.1(c),

Taking a moment about point O,

Then, $R_N(x) + Ft(a) = P(L)$

or $R_N(x) + \mu.R_N(a) = P(L)$

$R_N(x + \mu.a) = P(L)$

$$\text{Therefore,} \quad T_B = \frac{\mu PLr}{(x + \mu a)} \qquad (7)$$

Case 4:

Note that if the angle of contact is more than 60°, then non-uniform pressure occurs. Usually, the block or shoe is pivoted to the lever, instead of rigidly attached to the lever. This provides a uniform wear of the brake lining and wheel. The braking torque for a pivoted shoe brake (i.e. when $2\theta > 60°$) is

$$T_B = Ft.r = (\mu'.R_N)r$$

Where μ' is the equivalent coefficient of friction and is given by

$$\mu' = \frac{4\mu \sin\theta}{2\theta + \sin 2\theta}$$

These types of brakes provide more life and are also efficient in braking.

Self-energizing and self-locking brakes

The moment required for braking is the sum of moments of frictional force equal to $(\mu.R_N.a)$ and the moment of force (PL). As the moment due to braking force is supporting the moment of frictional force, the frictional force also helps to apply the brake. This type of brake is said to be a self-energizing brake. When the frictional force is great enough to apply the brake with no external force, then the brake is said to be the self-locking brake.

Consider the expression for braking torque, $T_B = \frac{\mu PLr}{(x + \mu a)}$ and in the denominator term can be positive if, $(x \leq \mu a)$, then the braking force applied P will be negative or equal to zero.

It shows that no external force is needed to apply the brake; hence, the brake is self-locking, and such condition is for the brake to be self-locking.

The self-locking brakes are used only in the back-stop rotation of applications. In general, the brake should not be self-locking but self-energizing. To avoid self-locking and to prevent the brake from self-locking, the distance x from the fulcrum should be kept greater than the value of (μa).

Example 1: The figure below shows a single block-shoe brake. The diameter of the wheel is 400 mm, and the angle of contact is 40°. The operating force of 500 N is applied at the end of a lever 600 mm long, and the position of the block is 250 from point O. The coefficient of friction between the drum and the lining is 0.35. Find the braking torque that may be transmitted by the block Brake.

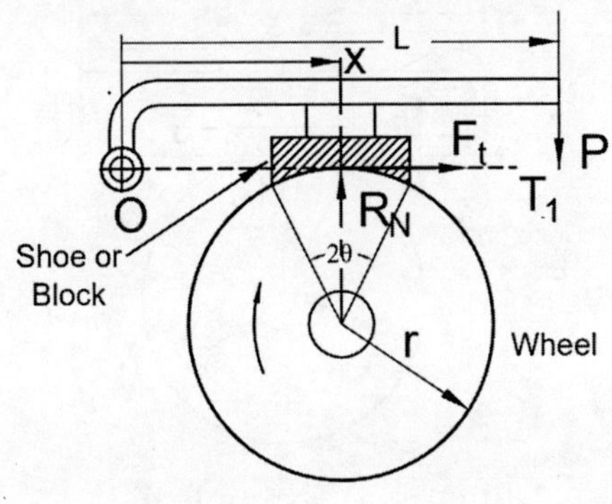

Fig.4.2 Example 4.1.

Solution

Given: D = 400 mm; therefore r = 0.2 m; 2θ = 40°; P = 500 N, L = 600 mm = 0.6 m

x = 250 mm = 0.25 m, μ = 0.35 (as the angle of contact is < 60°)

Taking moments about the fulcrum O, $R_N(x) = P(L)$

we get $R_N = 500\,(0.6/0.25) = 1200$ N

Braking torque, $T_B = (\mu.R_N).r$

$= 0.35 \times 1200 \times 0.2$

$= 84$ Nm

OR by using the equation, $T_B = (\mu.PL/x).r$ for braking torque when the Point O is in line with tangential to frictional force;

$$T_B = (\mu.PL/x).r$$
$$= (0.35 \times 500 \times 0.6/0.25) \times 0.2 = 84 \text{ Nm.}$$

Example 2: The figure below shows a single block-shoe brake. The fulcrum of the lever is 15 mm above the line of action of frictional force. The diameter of the wheel is 0.5 m, and the angle of contact is 50°. The operating force of 1 kN is applied at the end of a lever 0.750 m long and the position of the block is 0.25 from point O. The coefficient of friction between the drum and the lining is 0.35. Find the braking torque that may be transmitted by the block Brake when the wheel rotates clockwise and also counterclockwise.

Solution

Given: $a = 0.015$ m, $D = 0.5$ m; therefore $r = 0.25$ m; $2\theta = 50°$; $P = 1000$ N, $L = 0.75$ m $x = 0.25$ m, $\mu = 0.35$ (as the angle of contact is < 60°)

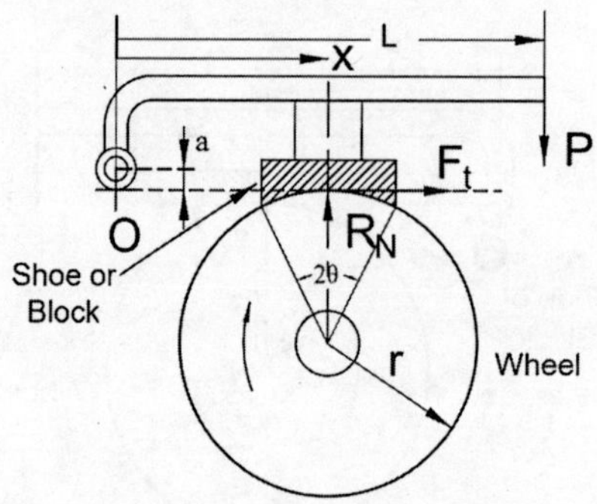

Fig.4.3(a) Example 4.2.

When the wheel is rotating clockwise, we use the equation,

$$T_B = \frac{\mu PLr}{(x + \mu a)}$$

$$T_B = \frac{0.35 \text{x } 1000 \text{ x } 1 \text{ x } 0.25}{(0.25 + 0.35 \text{ x}0.015)}$$

Braking torque required for clockwise direction = 342.8 Nm (1)

When the wheel is rotating counterclockwise, we use the equation,

$$T_B = \frac{\mu PLr}{(x - \mu a)}$$

$$T_B = \frac{0.35 \text{x } 1000 \text{ x } 1 \text{ x } 0.25}{(0.25 - 0.35 \text{ x}0.015)}$$

Braking torque required for counter clockwise direction = 357.5 m (2)

Example 3: The figure below shows a single block-shoe brake. The fulcrum of the lever is 20 mm below the line of action of frictional force. The diameter of the wheel is 0.6 m, and the angle of contact is 90°. The operating force of 700 N is applied at the end of a lever 0.8 m long, and the position of the block is 0.2 m from point O. The coefficient of friction between the drum and the lining is 0.35. Find the braking torque that may be transmitted by the block Brake when the wheel rotates clockwise and also counterclockwise.

Solution

Given: a = 0.02 m, D = 0.6 m; r = 0.3 m; 2θ = 90°; P = 700 N, L = 0.8 m,
x = 0.2 m, μ = 0.35.

as the angle of contact is 2 θ = 90° > 60°, then θ = 45° = 0.785 rad

The effective coefficient of friction μ' is given by

$$\mu' = \frac{4\mu \sin\theta}{2\theta + \sin 2\theta} = \frac{4x0.35x \sin 45^o}{2x0.785 + \sin 90^o} = 0.385$$

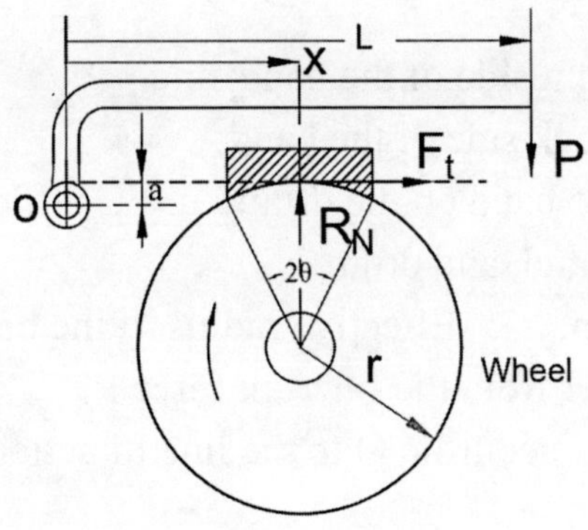

Fig.4.3(b) Example 4.3.

When the wheel is rotating clockwise, we use the equation,

$$T_B = \frac{\mu PLr}{(x - \mu a)}$$

$$T_B = \frac{0.385x\,700\,x\,0.8\,x\,0.3}{(0.2 - 0.385\,x0.02)}$$

Braking torque required for clockwise direction = 336.35 Nm (1)

When the wheel is rotating counterclockwise, we use the equation,

$$T_B = \frac{\mu PLr}{(x + \mu a)}$$

$$T_B = \frac{0.385x\ 700 \times 0.8 \times 0.3}{(0.2 + 0.385 \times 0.02)}$$

Braking torque required for counter-clockwise direction = 311.4 Nm (2)

4.5 Band brakes

A band brake consists of a drum partly covered with a flexible band of leather, or ropes, or steel lined with friction material. The band embraces the part of the circumference of the rotating drum, and when it is tightened up, the friction between them results in a retarded motion. A typical band brake is shown in Fig.4.4. In this simple band brake, one end of the band is attached to a fixed pin while the other end is attached to the lever at a distance from the fulcrum, as shown in the figure.

Let, T_1 = Tension in the tight side of the band, and

T_2 = Tension in the slack side of the band,

θ = Angle of lap of the band on the drum, μ = Coefficient of friction

t = Thickness of the band, and drum

r = Radius of the drum, r_e = Effective radius of the band = r + t/2

L = Length of the lever from the fulcrum force P

b = Perpendicular distance from O to the line of action of tensions

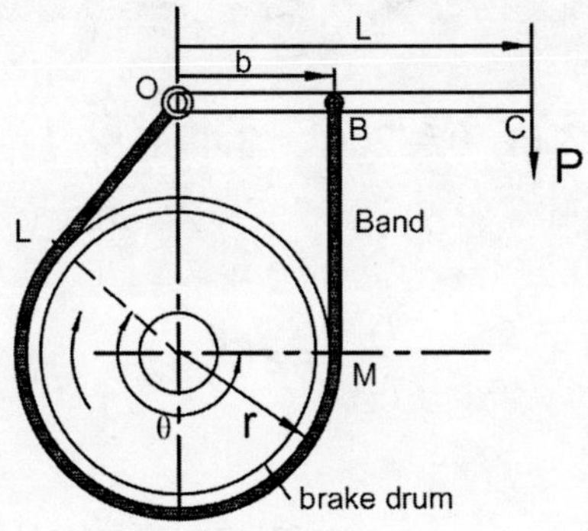

Fig.4.4 Band brake.

The limiting ratio of the tensions T_1 and T_2 is given by the relation,

$$\frac{T_1}{T_2} = e^{\mu\theta}$$

Braking torque on the drum is

$T_b = (T_1 - T_2)\, r$ for negligible band thickness

$T_b = (T_1 - T_2)\, r_e$ for band of thickness t

- When the brake drum rotates in the *clockwise direction*, the band attached to point B is tight side with tension T_1. The other end of band attached to the fulcrum O is slack side with tension T_2.
- On the other hand, when the drum rotates in the *anticlockwise direction*, the tensions in the band will be reversed,

Taking moments about the fulcrum O, we get,

$P.L = T_1.b$ (For clockwise rotation of the drum) and

$P.L = T_2.b$ (For anticlockwise rotation of the drum)

Exercise 4: A band brake acts on the circumference of a drum of 0.6 m diameter which covers 200°. This band brake produces a braking torque of 250 N-m. One end of the band is fixed to the fulcrum pin, and the other end to a pin 0.15 m from the fulcrum. The operating force is applied at 0.6 m from the fulcrum. The coefficient of friction is 0.25 for the brake material used.

Find the operating force when the drum rotates in the (a) anticlockwise direction, and (b) clockwise directions.

Solution

Given: d = 0.6 m or r = 0.3 m, θ = 200° = 3.49 rad, T_b = 250 Nm, b = 0.15 m, L = 0.6 m, μ = 0.25.

The ratio of Tensions in Band $$\frac{T_1}{T_2} = e^{\mu\theta}$$

or $$\frac{T_1}{T_2} = e^{0.25x3.49} = 2.39$$

The braking torque is given by $T_b = (T_1 - T_2)\, r$

or $(T_1 - T_2) = T_b/r$

$= 250/0.3 = 833.3$ Nm

Solving for T_1 and T_2

$T_1 = 1431.4$ N and $T_2 = 598.1$ N

Taking moments about the fulcrum O, we have for clockwise rotation of drum, Therefore

$P \times L = T_2 \times b$

$P = 598.1 \times 0.15/0.6$

$= 149.5$ N

Taking moments about the fulcrum O, we have for clockwise rotation of drum, Therefore

$P \times L = T_1 \times b$

$P = 1431.4 \times 0.15/0.6$

$= 347.6$ N

Exercise 5: A shaft carries a heavy flywheel having a mass of 600 kg. The normal speed of the shaft is 360 rpm and the flywheel has the effective radius of gyration of 0.55 m. The shaft is connected with a simple band brake whose drum diameter is 0.3 m. The angle of lap is 210°. The distance of brake force applied is 0.75 m and the end of band is hinged at a distance of 0.2 m from fulcrum. The coefficient of friction is 0.28 for the brake material. Determine the braking torque due to a hand load of 250 N. Also, calculate the number of turns of the wheel when the shaft is brought to rest.

Solution

Given: m = 600 kg, N = 360 rpm, or 28.27 rad/s, k = 0.55 m, d = 0.3 m, or r = 0.15 m, θ = 210° or 3.66 rad, L = 0.75 m, b = 0.2 m, μ = 0.28, P = 250 N.

The ratio of Tensions in Band $\frac{T_1}{T_2} = e^{\mu\theta}$

$$\frac{T_1}{T_2} = e^{0.28x3.66} = 2.78 \quad (1)$$

Taking moments about the fulcrum,

$$T_2 \times 0.2 = 250 \times 0.75$$
$$T_2 = 937.5 \text{ N}$$

Using above Eqn. (1),

$$T_1 = 2.78 \times 937.5$$
$$= 2606.2 \text{ N}$$

The braking torque due to these band tensions is

$$T_b = (T_1 - T_2)\, r$$
$$= (2606.2 - 937.5) \times 0.15$$
$$= 250.3 \text{ Nm.}$$

Number of turns required to flywheel come rest,

The kinetic energy of flywheel $= (1/2)\,[mk^2]\,\omega^2$

$$= (1/2) \times [600 \times 0.55^2]\, 28.27^2$$
$$= 72{,}526 \text{ Nm}$$

Above energy is utilized to overcome the work done due to the braking torque. Equating the kinetic energy of flywheel to the work done by the brake, we get

$$72{,}526 = 2\pi\,(n) \times 250.3$$

Number of turns = 289.76 say 290

4.6 Internal expanding shoe brakes

An internal expanding brake is the most popular among all the brakes. It works well for both directions of rotation of shafts in controlling speed through braking. It consists of a brake drum with an internal contact surface. The two independent split shoes with the liner rub against the contact surface of the brake drum for effecting braking. A conventional internal expanding brake system is shown in Fig.4.6. The outer surface of the shoes is lined with special friction material so as to increase the coefficient of friction and also to prevent the metal from wearing away. The shoes S_1 and S_2 are pivoted at one end, about a fixed fulcrum O_1 and O_2. The other end of the shoes is made to contact a cam.

The expansion of shoes is done by giving a displacement through the cam. As soon as the cam rotates, the shoes are moved outwards against the rim of the drum. The friction between the shoes' outer surface and the drum's inner surface produces the braking torque. The action of this led to retarding the rotation or the speed of the shaft connected.

During normal running, the braking system will not be effective or not operational. Whenever braking is required, the cam provided at the top is rotated slowly. This cam presses the shoe against the rotating drum resulting a rub between the rubbing surfaces. The action reduces the speed of the shaft driving. During normal running, the contact surfaces are held separately. The shoes are normally held in this off position by a spring as shown in figure. In internal expanding brake system, the drum encloses the entire mechanism, and is well protected from dust and moisture. These types of brakes are commonly used in motor cars and light trucks.

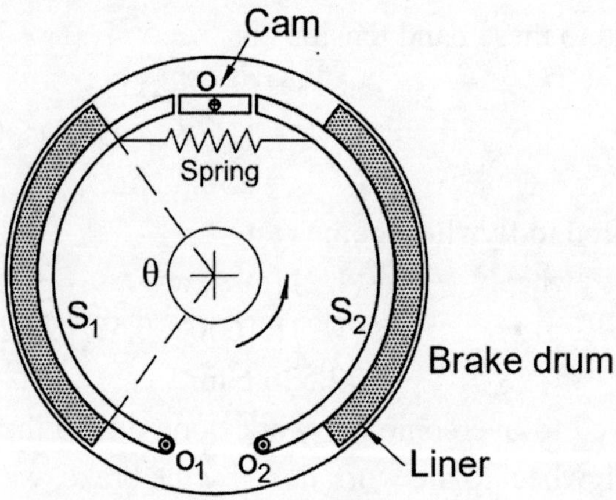

Fig.4.6 Internal expanding brake.

As shown in the figure above, assume that the drum is rotating in the anticlockwise direction. The frictional torque for the anticlockwise direction of the drum is opposite to the fulcrum O_1 of the left-hand shoe. This shoe is called as *leading* or primary shoe and the right-hand shoe is known astrailing or secondary shoe. When the drum rotates in a clockwise direction, the leading and astrailing shoes right and left shoes respectively.

The limitations of this braking system are that it tends to wear more and larger frictional contact surfaces are more susceptible to noise factors. This braking system suffers from heat dissipating issues and takes longer to stop.

4.7 Disc brakes

A disc brake is a popular mechanical system used to slow down or completely stop the motion of vehicles or rotating bodies. Disc brakes have become a fundamental component of modern automotive braking systems nowadays. These brakes are highly efficient braking mechanisms that have become significant due to their superior performance and safety features. Disc brakes offer improved stopping power, and the heat dissipation capacity compared to their older drum brake counterparts.

It consists of a flat, circular disc attached to the shaft or rotating wheel and a set of friction pads mounted on either side of the disc. To retard the motion, the friction pads are pressed against the rotating disc. The friction between the rotating disc and brake pads creates a frictional braking force to reduce the rotating body's speed. When the brake pedal is pressed, brake pads are squeezed against the disc, creating friction that converts kinetic energy into heat energy, ultimately slowing down or halting the vehicle's movement. Disc brakes are widely used in modern automotive safety and control systems. Because of their efficiency, heat dissipation, and improved braking performance, braking systems become crucial.

Working of Disc brakes: It work by converting kinetic energy into heat energy to slow down or stop a vehicle. Fig.4.7 shows the schematic diagram of the Disc brake. When the callipers squeeze the brake pads against the rotating disc, creating friction. This friction generates heat, which is then dissipated into the surrounding air, causing it to decelerate.

A disc brake consists of several components. The main elements include a rotor (disc), a calliper, brake pads, and a hydraulic system. A braking force is transferred to the master cylinder through a booster (servo system), which amplifies, converting it into hydraulic pressure. The braking force is applied on the right-side friction pad through the master cylinder. The pressure activates pistons within the brake callipers, which in turn force the brake pads, made of friction material, against the rotating brake rotors attached to the wheels. When the brake force is pressed, the calliper squeezes the brake pads against the rotor, creating friction that slows down or stops the vehicle.

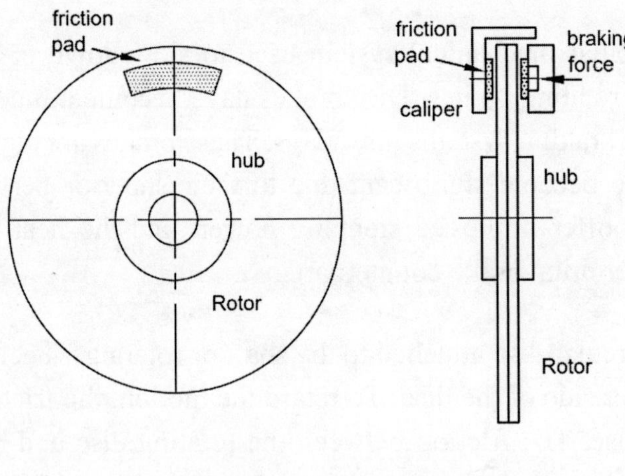

Fig.4.7 Disc brake.

4.8 Dynamometers

A dynamometer is a device for measuring mechanical power transmitted by a rotating shaft. It works like a brake but in addition, it has a device to measure the frictional resistance. By knowing the frictional resistance, the torque transmitted or power (product of torque and angular speed) of the engine can be estimated. Two types of dynamometers are used for measuring an engine's brake power: a) Absorption dynamometers and b) Transmission dynamometers.

In absorption-type dynamometers, all the energy produced by the engine or prime movers or any driving shaft is absorbed by the friction resistance of the brake system and transformed into heat energy during the process of measurement. In another type of dynamometer, called the transmission dynamometer, the energy is used for doing some work but not wasted in friction or heat. The energy or power produced by the engine is transmitted through the dynamometer to other machines where the power developed is suitably measured. Following are a few dynamometers used in the measurement of torque or power.

4.9 Rope brake dynamometer

It is an absorption type of dynamometer and is most commonly used for measuring the brake power of an engine. It is simple in construction and it has a pulley or like flywheel with one, or more ropes are wound around the flywheel or rim of a pulley. This pulley is fixed firmly to the shaft of an engine. The tension in the ropes' upper end is measured using a spring balance.

The other lower end of the rope has a provision to keep tension by applying a dead weight, as shown in Fig.4.8. In order to keep the rope in position on the pulley, guideways are provided and slipping the rope over the flywheel is avoided. Also, a few wooden blocks are used to keep the ropes in position at intervals around the circumference of the flywheel.

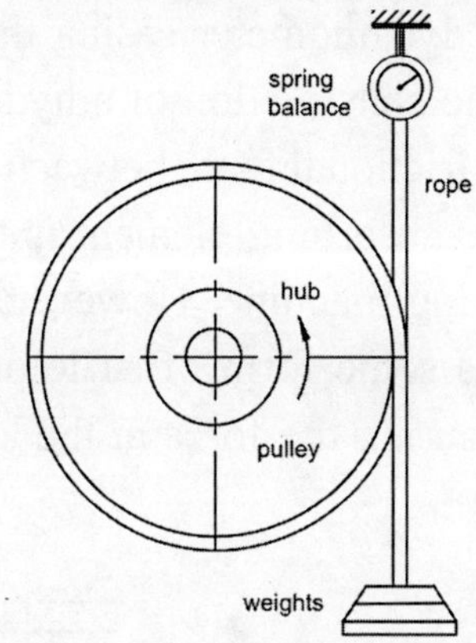

Fig.4.8 Rope brake dynamometer.

The frictional resistance of the brake is transformed into heat by the rope brake dynamometer. Therefore, the dynamometer is always cooled by circulating water inside the pulley. A provision is made in the pulley for circulating water continuously.

When the engine is running at a constant speed, the frictional torque, due to the rope, is equal to the torque being transmitted by the engine.

Let, W = Dead load in newtons, S = Spring balance reading in newtons, D = Diameter of the wheel in metres, d = diameter of rope in metres, and N = Speed of the engine shaft in rpm.

The net load on the brake $= (W-S)$

The distance moved in one revolution $= \pi\,(D+d)$

Work done per revolution $= (W-S) \times \pi\,(D+d)$

Work done per minute $= (W-S) \times \pi\,(D+d)\,N$

Therefore Power of engine $= (W-S) \times \pi\,(D+d)(N/60)$

From the above expression for power, power or torque can be determined by noting down the spring balance reading and the dead weight added to the rope at known constant speed of engine or rotational speed.

4.10 Hydraulic Dynamometer

The hydraulic dynamometer is used to measure the power of an engine or any prime movers. It is an absorption type of dynamometer, and it uses fluid friction for its operation to dissipate mechanical energy. The hydraulic dynamometer is also called a fluid friction dynamometer. Among all hydraulic dynamometers, Tesla fluid friction dynamometers and Froude water vortex dynamometers are popular. A hydraulic dynamometer has semicircular vanes placed in a rotor and stator. Water flows in a toroidal vortex around the vanes and creates a torque reaction in the dynamometer casing that is resisted by the dynamometer and measured by a load cell. The construction of a hydraulic dynamometer is similar to a fluid flywheel that measures the frictional force between impeller vanes and a moving fluid. In this dynamometer, the power-absorbing element is the housing, which tends to rotate with the input shaft of the driving machine. However, such rotation is constrained by a force-measuring device, such as some form of scale or load cell, placed at the end of a reaction arm of radius. By measuring the force at the known radius, the torque T may be computed by the simple relation.

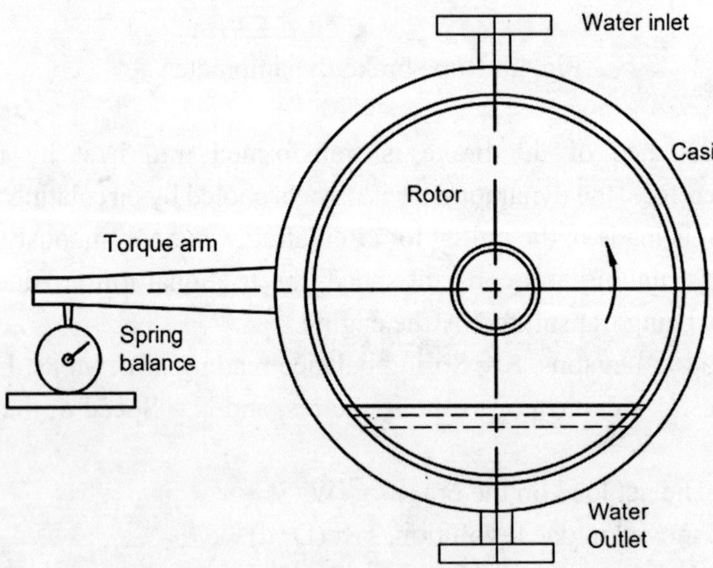

Fig.4.9 Hydraulic dynamometer.

A Tesla fluid friction dynamometer consists of a disc connected to a revolving shaft. The dynamometer is completely enclosed in stationary casting. The space between the outer casing and the revolving disc is partly filled with water or any recommended viscous fluid.

The casing of the dynamometer is attached to one end of the lever and the other end, which balances the weight. The relative motion of the disc and casing balances the fluid frictional and viscous forces used for braking action. The torque of the engine and the torque developed due to viscous forces are balanced by the weight at known leverage and are used to assess the engine power. This arrangement is suitable for power measurement when speed is high, less torque and viscous forces are also small. Fig.4.9 shows a hydraulic dynamometer in its simplest form which acts as a water brake in measuring power.

The Vortex fluid friction dynamometer consists of a rotating disk carried with several semi-elliptical blades forming brake chambers. The disc is connected to the driving shaft of the engine or primary mover whose power is to be estimated. The disc revolves inside the stationary casing that carries the counter blades projecting from the inner wall. These will form another part of the working brake chamber. The runner and the casing are filled with water and flow continuously while testing. The casing of the dynamometer is attached to one end of the lever and the other end, which balances the weight. As the disc rotates, the Vortex and the Eddy currents are set up in the water. The mechanical power is then transformed into fluid friction in the form of a rising temperature of the fluid. The capacity of the dynamometer to measure varied powers depends on fluid friction that can be carried out. This type of dynamometer is used for a wide range of power and speed. Hydraulic dynamometers are well suited for testing high load capacity and have the ability to handle heavy machinery and large engines. These are commonly used in industries involved in the testing of large engines, turbines, pumps, etc. This type of dynamometer is considered for larger capacities than any simple mechanical brake because of the heat generated. This dynamometer type easily removes the heat developed during the test by circulating the water into and out of the brake chamber.

4.11 Eddy Current Dynamometer

An eddy current dynamometer is an electromechanical energy conversion device that converts mechanical energy to electrical energy. The fundamental working principle of this dynamometer is Faraday's Law of electromagnetic induction. Eddy current dynamometer works on the principle that when a conductor of rotor moves through a magnetic field, it induces eddy currents. These eddy currents flow in a short circular path around the conductor and are dissipated in the form of heat.

It consists of a stator on which are fitted a number of electromagnets and a rotor disc made of copper or steel and coupled to the engine's output shaft. When the rotor rotates, eddy currents are produced in the stator due to magnetic flux set up by the passage of field current in the electromagnets. These eddy currents oppose the motion of the rotor, thus loading the engine. The eddy currents are dissipated in producing heat, so this type of dynamometer also requires some cooling arrangements. The torque is measured similarly to absorption dynamometers, i.e., with the moment arm's help. The load is controlled by regulating the current in the electromagnets. A schematic of the Eddy current dynamometer is shown in Fig.4.10

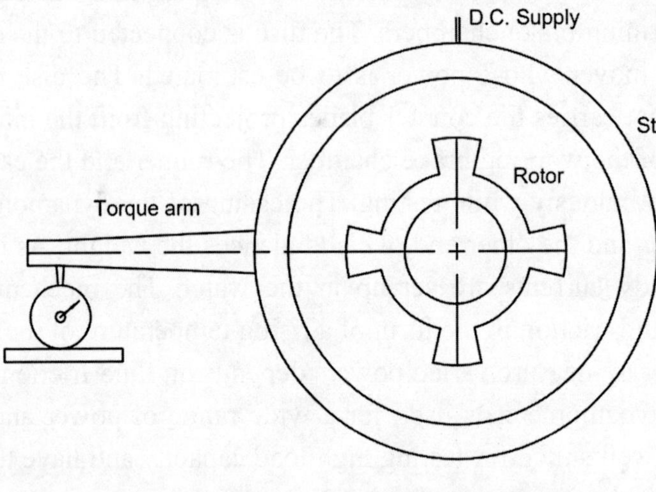

Fig.4.10 Eddy current dynamometer.

An eddy current dynamometer is compact and compatible with the other prime movers. The advantages of this dynamometer are that it has fewer losses and is highly efficient. When the dynamometer is coupled with the engine, the rotor of the dynamometer cuts the stator magnetic field, and an EMF is induced on the rotor conductors. This will produce the eddy current to flow in the rotor conductors. The direction of the eddy currents is opposite to the change in the magnetic flux and is generated in the rotor. The force is exerted by the magnetic flux is opposed by the rotor. Due to the prime mover's continuous input, it keeps rotating. The arm of the eddy current dynamometer is connected to the body of the stator. The position of the end of the arm can measure the torque produced in the rotor winding. The power transmitted from the engine can be estimated using other specifications like the rotor's speed.

A clutch is a mechanical device used to connect or disconnect the rotary motion or power from sources like engines, electric motors, etc., to the applications or machinery. The presence of a clutch in motion and power transmission is very important when the motion has to be stopped or started frequently. Otherwise, the clutch controls the flow of mechanical power when a common motor, engine or turbine drives multiple shafts. The cutches are in operation whenever power is required at the operators' will. There are two types of clutches: positive clutches and friction clutches.

4.12 Clutches

Positive clutches are used to engage or disengage the power from the motor running at low speeds. Common positive clutches have engaging surfaces like jaw-type, and square jaw type or spiral jaw type are in use. Friction clutches work on the principle of the frictional forces developed when two or more surfaces are in contact. Friction clutches are usually very popular over the jaw clutches due to their better performance. Friction surfaces can be used at high engagement speeds and slip during the engagement, which enables the smother pickup to accelerate with less shock. The friction clutches are found in cases in which power is to be delivered to machines partially or fully loaded. In automobiles, a friction clutch is used to connect the engine to the driving wheel. The friction clutches engage easily and gradually pick up the driven shaft up to proper speed.
The popular friction clutches that are in use are: 1. Plate or Disc clutches (single disc or multiplate disc clutch), 2. Centrifugal clutches. and 3. Cone clutches.

4.13 Single plate clutch

A single disc or plate clutch frequently engages or disengages the driving shaft with small powers. In construction, it has a clutch plate coated on both sides with a friction material (Ferrodo). It can freely slide axially with splines of the shaft. The pressure plate is used to apply pressure to the clutch plate for power transmission and is mounted inside the clutch body, which is bolted to the flywheel. Both the pressure plate and the flywheel rotate with the driving shaft. The pressure plate pushes the clutch plate towards the flywheel by a set of strong springs arranged radially inside the body. A schematic layout of the components of a single plate clutch is shown in Fig.4.11(a).
The disengaging of the power from the driving shaft using the plate clutch is done by applying force to move the pressure plate away from the flywheel. This action removes the pressure from the clutch plate and thus moves away from the flywheel, resulting in the driven shaft becoming stationary. On the other hand, if the force is withdrawn, the springs extend, and thus, the pressure plate pushes the clutch plate back towards the flywheel. Then the driving shaft engages with the driven also motion and power is transmitted as usual.
A single disc or plate clutch frequently engages or disengages the driving shaft with small powers. In construction, it has a clutch plate coated on both sides with a friction material

(Ferrodo). It can freely slide axially with splines of the shaft. The pressure plate is used to apply pressure to the clutch plate for power transmission and is mounted inside the clutch body, which is bolted to the flywheel. Both the pressure plate and the flywheel rotate with the driving shaft. The pressure plate pushes the clutch plate towards the flywheel by a set of strong springs arranged radially inside the body. A schematic layout of the components of a single plate clutch is shown in Fig.4.11(b).

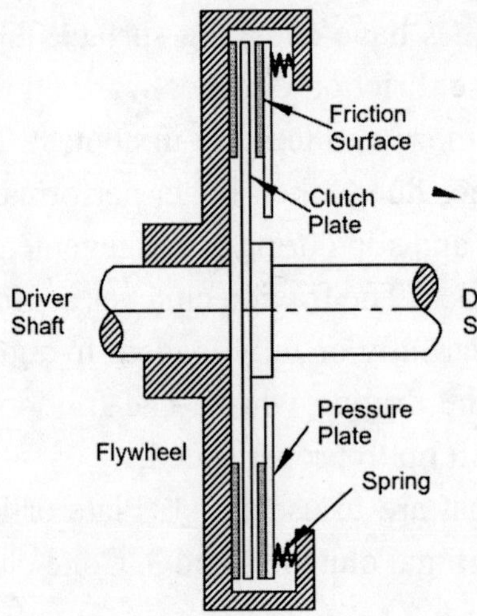

Fig.4.11(a) Single plate clutch.

The disengaging of the power from the driving shaft using the plate clutch is done by applying force to move the pressure plate away from the flywheel. This action removes the pressure from the clutch plate and thus moves away from the flywheel, resulting in the driven shaft becoming stationary.

On the other hand, if the force is withdrawn, the springs extend, and thus, the pressure plate pushes the clutch plate back towards the flywheel. Then the driving shaft engages with the driven also motion and power is transmitted as usual.

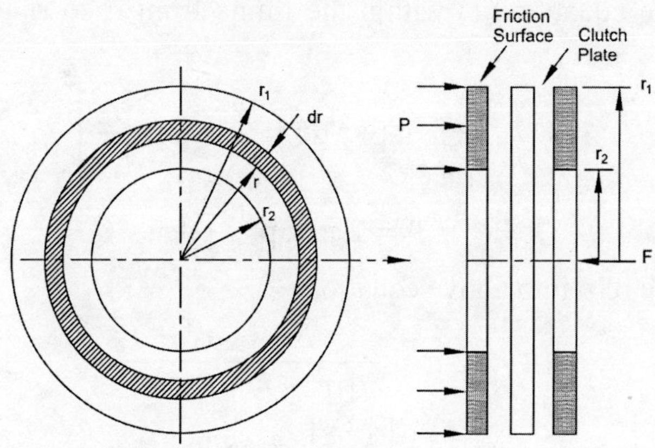

Fig.4.11(b) Single plate clutch.

Now consider two friction surfaces between the clutch plate as shown above and is maintained in contact by an axial thrust F

Let T = Torque transmitted by the clutch,

p = Intensity of axial pressure required to hold contact surfaces together,

r_1 and r_2 = External and internal radii of friction faces and

μ = Coefficient of friction.

F = Axial thrust required to keep frictional surfaces together.

Consider an elementary ring of radius r and thickness dr as shown in Fig. 4.8(b).

The area of contact friction surface, $= (2\,\pi\, r).dr$

∴ Normal or axial force on the ring, δF = Pressure × Area

$= (p \times 2\,\pi\, r).dr$

Above frictional force on the ring acting tangentially at radius r,

$$F_r = \mu.\delta W = \mu.p \times (2\,\pi\, r).dr$$

∴ Frictional torque acting on the ring, $T_r = F_r \times r = \mu.p \times 2\,\pi\, r.dr \times r$

$$= 2\,\pi \times \mu\,.p.r^2\, dr \qquad (8)$$

Case 1: Torque transmission by uniform pressure

the pressure is uniformly distributed over the entire area of the friction face, then the intensity of pressure over this area is

$$p = \frac{F}{\pi[r_1^2 - r_2^2]}$$

Integrating the above equation (1) within the limits from r_2 to r_1 for the total frictional torque.

$$T = \int_{r_1}^{r_2} 2\,\pi\;\mu\,pr^2\,dr = 2\,\pi\;\mu\,p\left[\frac{r^3}{3}\right]_{r_1}^{r_2}$$

Or
$$T = 2\,\pi\;\mu\,p\,\frac{r_1^3 - r_2^3}{3}$$

Substituting the value of p into above equation

$$T = 2\,\pi\;\mu\,\frac{F}{\pi[r_1^2 - r_2^2]}\,\frac{r_1^3 - r_2^3}{3}$$

$$T = \frac{2}{3}\mu\,F\,\frac{[r_1^3 - r_2^3]}{[r_1^2 - r_2^2]} = \mu FR \qquad (9)$$

Where R is the mean radius of friction surface for a case of uniform pressure

$$R = \frac{2\,[r_1^3 - r_2^3]}{3\,[r_1^2 - r_2^2]}$$

Case 2: Torque transmission by uniform wear

Considering the uniform wear of frictional surface, the intensity of pressure 'p'varies inversely with the distance r from the centre. Therefore p.r. = C (a constant)
The normal force on the ring of width dx is, δF = (2πr)p.dr and by substituting p.r = C
We get
$$\delta F = 2\pi\,C.dr$$

Total force acting on the friction surface is

$$F = \int_{r_1}^{r_2} 2\,\pi\;C\,dr = 2\,\pi\,C\,(r_1 - r_2)$$

Therefore
$$C = \frac{F}{2\,\pi\,(r_1 - r_2)}$$

Then, the frictional torque is given by $T_r = 2\pi\mu pr^2 dr = 2\pi\mu Cr.\,dr$

The total frictional torque on the friction surface

$$T = \int_{r_1}^{r_2} 2\pi\mu Cr.\,dr = 2\pi\mu C\frac{(r_1^2 - r_2^2)}{2}$$

$$T = \pi\mu\,\frac{F(r_1^2 - r_2^2)}{2\,\pi\,(r_1 - r_2)} = \mu F\,\frac{(r_1 + r_2)}{2} \qquad (10)$$

Total frictional torque on the friction surface **T= μFR**

Where R is the mean radius equal to $\frac{(r_1+r_2)}{2}$

In general, total frictional torque acting on the friction surface (or on the clutch) is given by

$$T = n.\mu.F.R$$

Where n = Number of pairs of friction or contact surfaces,

R = Mean radius of friction surface

For uniform pressure $R = \frac{2}{3}\frac{[r_1^3 - r_2^3]}{[r_1^2 - r_2^2]}$ And for uniform wear $R = \frac{(r_1+r_2)}{2}$

Note:

- In a single disc or plate clutch, both sides of the disc are effective. Therefore, a single it has two pairs of surfaces in contact, i.e. n = 2.
- For a new clutch, the intensity of pressure is uniform but in an old clutch the uniform wear theory is more suitable.
- The uniform pressure theory gives a higher frictional torque than the uniform wear theory. Therefore, uniform wear should be considered unless otherwise stated.

4.14 Multiplate clutch

A multiplate clutch is used where a large torque is to be transmitted. The inside discs are fastened to the driven shaft and slide along the shaft. Fig.4.12 shows the schematic diagram of the multiple plate clutch. The outside discs are held by bolts and fastened to the housing keyed to the driving shaft. The multiplate clutches are extensively used in motor cars, machine tools, etc. The driving discs are pressed against the driven discs by a suitable mechanism, and the torque is transmitted by friction between the discs.

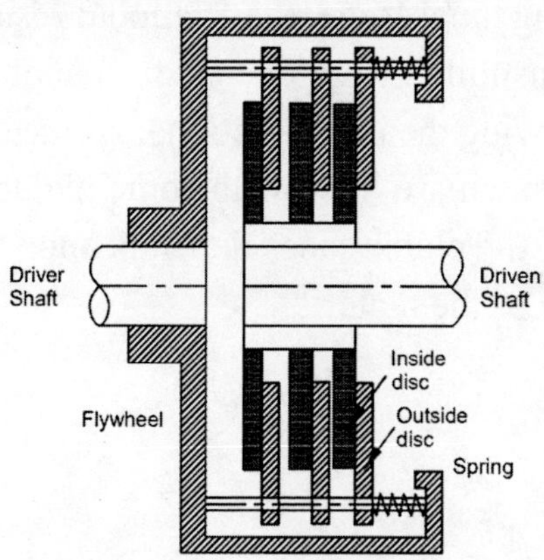

Fig.4.12 Multiplate clutch.

Let n_1 and n_2 be the number of discs on the driving shaft and the driven shaft. Then Number of pairs of contact surfaces,

$$n = n_1 + n_2 - 1$$

Total frictional torque acting on the friction surfaces or on the clutch,

$$T = n.\mu.F.R$$

R = Mean radius of the friction surfaces

For uniform pressure, $R = \frac{2}{3}\frac{[r_1^3 - r_2^3]}{[r_1^2 - r_2^2]}$

For uniform wear, $R = \frac{(r_1 + r_2)}{2}$

4.15 Cone clutch

It consists of one pair of friction surfaces only and is used where a small torque is to be transmitted. The frictional contact surface is a frustum cone; therefore, the plane of contact is inclined to the axis of the shaft. The driver is keyed to the clutch housing of the cone clutch and forms the inner frictional surface, as shown in the figure. This inner conical face exactly fits into the outer conical frictional surface of the driven shaft.

The driven member of the clutch is keyed to the driven shaft and can be shifted along the shaft by a forked lever. A spring pressure is used on the driven shaft so that the contact is maintained always in order to engage the clutch by bringing the two conical surfaces in contact. Because of this spring axial force, the frictional resistance is set up at this contact surface, and the torque is transmitted from one shaft to another. The disengagement of the clutch is achieved by withdrawing the inner frictional surface axially along the driven shaft. The contact surfaces of the clutch are lined with some material like wood, leather, cork or asbestos, etc. The material of the clutch faces depends upon the allowable normal pressure and the coefficient of friction of the material.

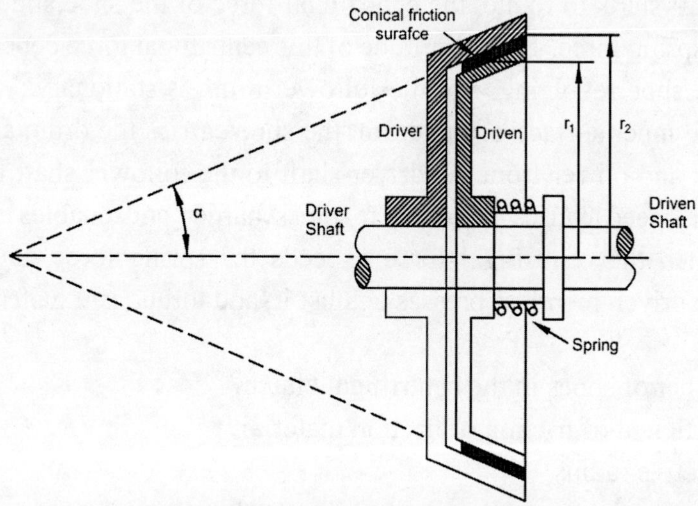

Fig.4.13 Cone clutch.

Consider a pair of friction surfaces as shown in Fig.4.13. The area of contact of a pair of friction surfaces is a frustum of a cone. Therefore, the torque transmitted by the clutch is given by

$$T = \frac{2}{3}\mu F.cosec(\alpha)\frac{[r_1^3 - r_2^3]}{[r_1^2 - r_2^2]} \quad \text{for uniform pressure,} \quad (11)$$

$$T = \mu F.cosec(\alpha)\frac{(r_1 + r_2)}{2} \quad \text{for uniform wear,} \quad (12)$$

Where, r_1 and r_2 are the outer and inner radius of friction surfaces respectively.
α = Semi angle of the cone
μ = Coefficient of friction between contact surfaces,
F = braking force applied.

4.16 Centrifugal clutch

The centrifugal clutches are used to transmit power from the motor drive to the follower above the specific and minimum speed. It consists of a number of shoes inside the clutch and freely moves radially outward with the centrifugal force developed during the rotation. These shoes are held against the boss on the driving shaft by means of springs. The outer surface of the shoes is covered with a friction material. The centrifugal clutch is shown in Fig.4.14.

As the drive shaft starts to rotate, the centrifugal force of the shoe slides outward radially and against the spring force. The magnitude of this centrifugal force depends upon the speed and mass of the shoe revolving. But the follower drum is stationary. When the revolving shoe touches the inner surface of the drum, the shoe carries the drum along with it. Thus, carrying motion and power from the driver shaft to the follower shaft through the clutch. The increase of speed causes the shoe to press harder and enables more torque to be transmitted. When the centrifugal force exceeds the spring force, the shoe comes into contact with the driven member, presses against it, and torque is transferred.

Let n = number of shoes in the centrifugal clutch,
μ = coefficient of friction of friction material
R = effective radius
P_c= centrifugal force acting on each shoe at the running speed.
P_s= the inward force on each shoe at which engagement begins

Let the centrifugal force act on each shoe at the running speed is given by

$$P_c = m.\omega^2.r$$

the inward force on each shoe exerted by the spring at the speed N_1 at which engagement begins to take place,

$$P_s = m\,(\omega_1)^2\,r$$

Therefore, the net outward radial when the shoe presses against the rim = $(P_c - P_s)$ and the frictional force acting tangentially on each shoe, $F = \mu\,(P_c - P_s)$
Frictional torque acting on each shoe,

$$T = F \times R = \mu\,(P_c - P_s)\,R$$

Total frictional torque transmitted with n number of shoes of the centrifugal clutch is

$$T = n.\,\mu\,(P_c - P_s)\,R \qquad (13)$$

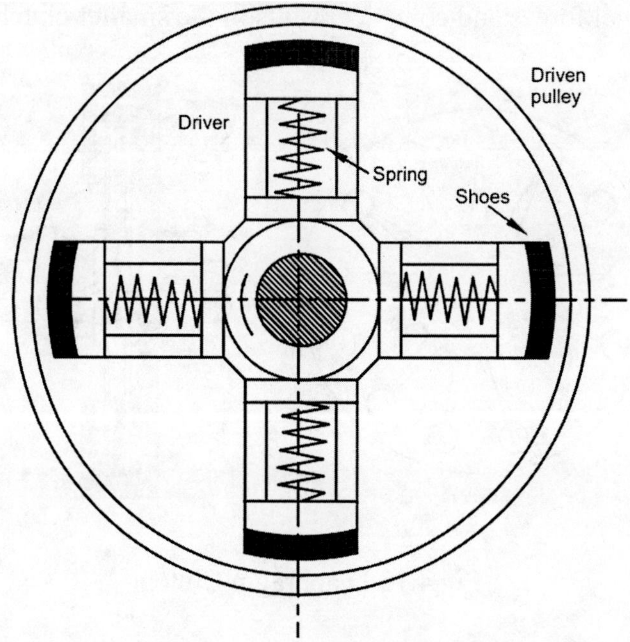

Fig.4.14 Centrifugal Clutch.

4.17 Diaphragm Clutch

In this type of clutch, a metallic diaphragm is used as a spring instead of coil or helical springs. This diaphragm always presses against the pressure plate to keep contact with the disc while transferring the torque from the driver to the driven shafts. While disengaging, the lever presses the diaphragm to become flat. The central position of the diaphragm spring is divided into several segments by radial slots. These segments act like springs, providing the required thrust on the pressure plate.

The leverage provided in the clutch for the diaphragm pushes the pressure plate back. This results in a gap between the disk and the pressure plate. Therefore, no power or motion is transferred until the braking force is withdrawn. This required braking force is less than that required by coil spring clutches.

The clutch consists of a diaphragm type of spring, conventional friction clutch, thrust plate, and release sleeve. The arrangement of this clutch is shown in Fig.4.15. In the engaged position, the diaphragm spring pivots on the inner pivot ring as shown. The diaphragm is held on the clutch cover, and the outer rings press the pressure plate. In this conical position, the diaphragm exerts pressure to keep the pressure plate in contact with the clutch plate and flywheel. The advantages of the diaphragm clutch are that the diaphragm spring is less affected by centrifugal forces and compact results in the smaller clutch housing.

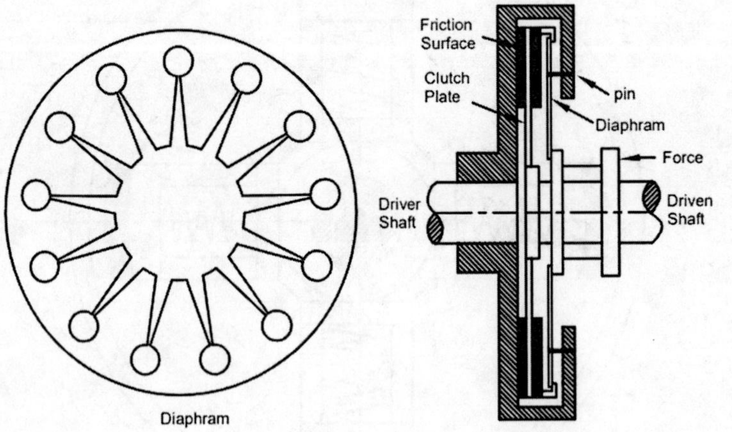

Fig.4.15 Diaphragm Clutch.

Example 6: A single plate clutch to connect an engine with the driven shaft has outer and inner diameters of 300 mm and 200 mm, respectively. The maximum intensity of pressure at any point in the contact surface is not to exceed 0.1 N/mm^2. Both sides of the clutch plate are effective, and the coefficient of friction is assumed to be 0.3. Determine the power transmitted by the clutch at a speed of 1500 rpm.

Solution:

Given:

$d_1 = 250$ mm; or $r_1 = 0.125$ m $\quad$ $d_2 = 200$ mm; or $r_2 = 0.1$ m

$p = 0.1$ N/mm^2; or 0.1×10^6 N/m^2 $\quad$ $n = 2$; $\mu = 0.3$

$$\text{Speed of clutch } \omega = \frac{2\pi N}{60} = \frac{2\,\pi \times 1500}{60} = 157.1 \text{ rad/s}$$

Consider a case of uniform wear, and therefore, the intensity of pressure p is maximum at the inner radius (r_2)

The axial thrust is given by $\quad F = 2\,\pi\, p.\, r_2\,(r_1 - r_2)$

$$= 2\,\pi \times (0.1 \times 10^6)\, 0.1\, (0.125 - 0.1) = 1571 \text{ N}$$

The mean radius of the friction surfaces for uniform wear,

$$R = \frac{r_1 + r_2}{2} = \frac{0.125 + 0.1}{2} = 0.1125\, m$$

We know that the torque transmitted,

$$T = n.\mu.W.R$$
$$= 2 \times 0.3 \times 1571 \times 0.1125$$
$$= 106 \text{ N-m}$$

∴ Power transmitted by a clutch,

$$P = T.\omega = 106 \times 157.1$$
$$= 16652 \text{ W} \quad \text{or} \quad 16.65 \text{ kW}$$

Example 7: A multiple-disc clutch has four pairs of working friction surfaces. The intensity of pressure on the friction surface is not to exceed 0.13 N/mm^2. Find the power transmitted when the clutch is running at 750 rpm. The outer and inner radii of friction surfaces are 250 mm and 175 mm respectively. Assume uniform wear and assume the coefficient of friction as 0.3.

Solution:

Given:

$p = 0.13$ N/mm^2; or 0.13×10^6 N/m^2

N = 750 rpm; therefore Speed $\omega = \frac{2\pi N}{60} = \frac{2\,\pi \times 750}{60} = 78.55$ rad/s

$d_1 = 250$ mm; or $r_1 = 0.125$ m $\quad d_2 = 175$ mm; or $r_2 = 0.0875$ m

$n = 4$; $\mu = 0.3$

Consider a case of uniform wear, and therefore, the intensity of pressure p is maximum at the inner radius (r_2)

The axial thrust is given by $F = 2\,\pi\,p.\,r_2\,(r_1 - r_2)$

$$= 2\,\pi \times (0.13 \times 10^6)\, 0.0875\, (0.125 - 0.0875)$$
$$= 2680.5 \text{ N}$$

The mean radius of the friction surfaces for uniform wear,

$$R = \frac{r_1 + r_2}{2} = \frac{0.125 + 0.0875}{2} = 0.10625\,m$$

We know that the torque transmitted,

$$T = n.\mu.W.R$$
$$= 4 \times 0.3 \times 2680 \times 0.10625$$
$$= 341.7 \text{ N-m}$$

∴ Power transmitted by a clutch,

$$P = T.\omega = 341.7 \times 78.55$$
$$= 26840 \text{ W} \quad \text{or} \quad 26.84 \text{ kW}$$

Example 8: A plate clutch is used between the driver and driven to transmit 10 kW at 1440 rpm and has four pairs of contact surfaces. Three discs are on the driving shaft, and two discs are on the driven shaft. The contact surfaces have an outside diameter of 280 mm and an inside diameter of 200 mm. Assuming the coefficient of friction as μ = 0.3, Find the total spring load required to keep the plates together, with the cases, a) uniform pressure and b) uniform wear.

Solution:

Given: P= 10 kW = 10 x10^3 W;

N = 1440 rpm; Speed $\omega = \frac{2\pi N}{60} = \frac{2\,\pi \times 1440}{60} = 150.8\ \text{rad/s}$

d_1 = 280 mm; or r_1 = 0.14 m; d_2 = 200 mm; or r_2 = 0.1 m; n = 4; μ = 0.3.

The power transmitted by the clutch P = T ω

Therefore the torque on the clutch T = P/ ω

= 10 x10^3/ 150.8 = 66.313 Nm

a) Mean radius R of the contact surface, for uniform pressure,

$$R = \frac{2}{3}\frac{[r_1^3 - r_2^3]}{[r_1^2 - r_2^2]}$$

$$= \frac{2}{3}\frac{[0.14^3 - 0.1^3]}{[0.14^2 - 0.1^2]} = 0.121 \text{ m}$$

Total frictional torque acting on the friction surfaces or on the clutch,

T = n.μ.F.R

Therefore, the total spring load required

$$F = \frac{T}{n.\mu..R} = \frac{66.313}{4 \times 0.3 \times 0.121} = 456.7 \text{ N}$$

b) Mean radius R of the contact surface, for uniform pressure,

$$R = \frac{r_1 + r_2}{2} = \frac{0.14 + 0.1}{2} = 0.12\ m$$

Therefore, the total spring load required

$$F = \frac{T}{n.\mu..R} = \frac{66.313}{4 \times 0.3 \times 0.12} = 460.5 \text{ N}$$

4.18 Bearings

The role of a bearing, in general, is to support and guide other moving members, such as rotating shafts. The bearing permits fixed direction motion between two parts, typically rotation or linear movement in a constrained way. Bearings that support the rotating shafts may be grouped by the direction in which the load acts on them. A rotating shaft may be subjected to radial and axial loads. Radial bearings or journals bear the radial loads. The rotating shafts are also frequently subjected to axial loads or thrust. The shafts carrying propellers in ships, helical gears, steam turbines, and vertical machine shafts are examples of thrust loadings. The pivot and collar bearings support the axial thrust of the rotating shafts.

The pivot provides the bearing surfaces at the end of a shaft to take the axial thrust. A collar bearing supports the axial thrust and can be conveniently placed anywhere along the axis of the shaft. The pivot surface may be a flat or conical surface, as shown in Fig.4.16(a) and (b). The collar bearing may have a flat surface with a single collar or several collars along the shaft length. The Fig. 4.16 (c) shows a single collar thrust bearing.

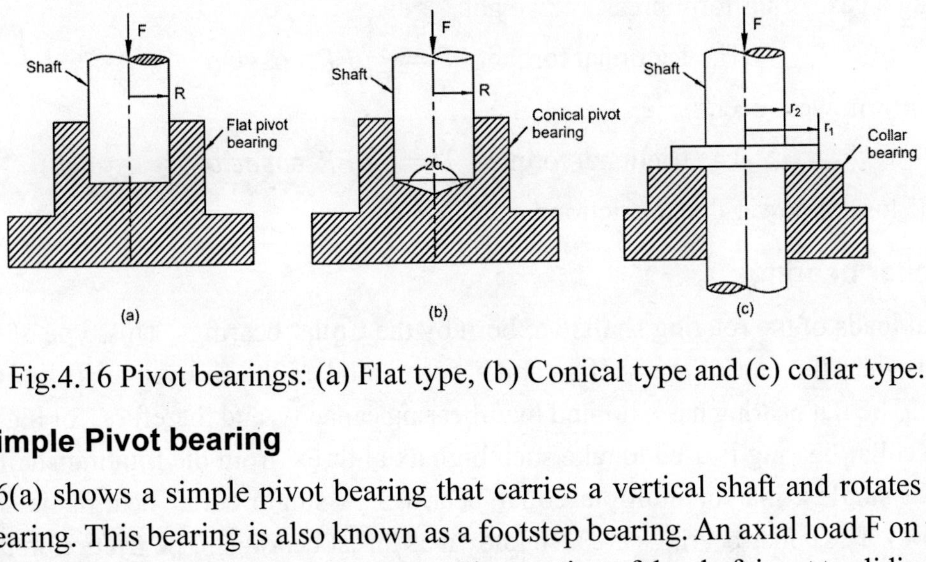

Fig.4.16 Pivot bearings: (a) Flat type, (b) Conical type and (c) collar type.

4.19 Simple Pivot bearing

Fig.4.16(a) shows a simple pivot bearing that carries a vertical shaft and rotates in a flat pivot bearing. This bearing is also known as a footstep bearing. An axial load F on the shaft results in a pressure over the area of contact. The rotation of the shaft is set to sliding friction along the surface of contact between the shaft and the bearing. The friction between the contact surfaces results in a loss of power.

The intensity of pressure per unit area of the bearing surface that is rubbing the surface between the shaft and bearing is $F/(\pi R^2)$, where R is the radius of the bearing surface. Assuming a case of uniform pressure, we get

The frictional torquc $T = \frac{2}{3}\ \mu FR$

When uniform wear occurs,

The frictional torque $T = \frac{1}{2}\ \mu FR$ (14)

Then, the loss of power due to friction is $\boldsymbol{P = T\,\omega}$

4.20 Conical Pivot Bearing

A conical pivot bearing that carries a vertical shaft is shown in Fig.4.16(b) and rotates in a conical-shaped pivot support bearing. This bearing surface is larger compared to a flat pivot bearing. An axial load F on the shaft results in a pressure over the area of contact. The rotation of the shaft is set to sliding friction along the surface of contact between the shaft and the bearing. The included angle of contact of the bearing surface is (2α), as shown in the figure. The intensity of pressure per unit area of the bearing surface that is rubbing the surface between the shaft and bearing. The friction between the surfaces results in a loss of power.

Assuming a case of uniform pressure, we get

The frictional torque $T = \frac{2}{3}\ \mu FR.\,cosec\ \alpha$

When uniform wear occurs,

The frictional torque $T = \frac{1}{2}\ \mu FR.\,cosec\ \alpha$ (15)

Then, the loss of power due to friction is $\boldsymbol{P = T\,\omega}$

4.21 Collar bearings

The axial loads of the rotating shafts are born by the Collar bearings. This type of bearings is useful when the end of the shafts cannot accommodate the bearing at its end like flat bearings. The flat bearing has a limited load-bearing capacity, and therefore, for higher axial thrust, a collar bearing is used to take such high axial thrust from the rotating shafts. There may be a single collar or multiple collar bearings. A single collar bearing is shown in Fig.4.16(c). The collar bearings are also known as thrust bearings. The friction in the collar bearings results in loss of power due to friction. Let r_1 and r_2 be the radius of the collar and the shaft, respectively. The frictional torque and the power loss due to the friction are as follows.

Assuming a case of uniform pressure, we get

The frictional torque $T = \frac{2}{3}\,\mu\,F\frac{[r_1^3 - r_2^3]}{[r_1^2 - r_2^2]}$

When uniform wear occurs,

The frictional torque $T = \mu F\frac{(r_1+r_2)}{2}$

Then, the loss of power due to friction is $\boldsymbol{P = T\omega}$

Where ω is the rotational speed of shaft in rad/s.

In case of multiple collars, the bearing pressure is distributed over all the collars and number of collars (n) required is calculated using the following equation. The Fig.4.17 shows the multiple collars bearing.

$$p = \frac{F}{n\pi(r_1^2 - r_1^2)}$$

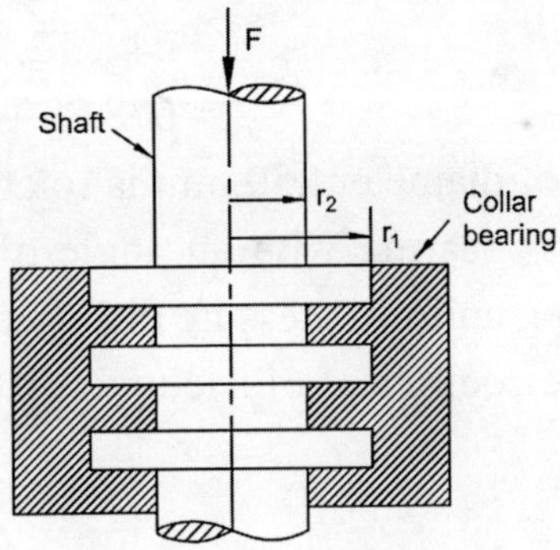

Fig.4.17 Multiple collars bearing.

Example 9: A vertical shaft of a compressor of diameter 200 mm is rotating at 120 rpm. The compressor shaft is supported on a flat footstep bearing. The shaft carries a vertical load of 24 kN. Assuming uniform pressure distribution and wear, estimate the power lost due to friction. Assume the coefficient of friction as 0.045.

Solution

Given:

D = 200 mm, R = 0.1 m, N = 120 rpm, F = 24 x 10^3 N, μ = 0.045

Speed of the shaft ω = 2 π N/60 = (2 π x120)/60 = 12.57 rad/s

a) Assuming a uniform pressure distribution,

The total frictional torque $T = \frac{2}{3}\mu F R$

$$= \frac{2}{3} \text{ x } 0.045 \text{ x } 24\text{x}10^3 \text{ x } 0.1$$

$$= 72 \text{ Nm}$$

The loss of power due to friction is $P = T\,\omega$

$= 72 \text{ x } 12.57$

$= 905 \text{ W}$

b) Assuming a uniform wear

The total frictional torque $T = \frac{1}{2}\,\mu F R$

$= \frac{1}{2} \text{ x } 0.045 \text{ x } 24\text{x}10^3 \text{ x } 0.1$

$= 54 \text{ Nm}$

The loss of power due to friction is $P = T\,\omega$

$= 54 \text{ x } 12.57$

$= 678.78 \text{ W}$

Example 10: A vertical shaft of diameter 250 mm is rotating at 240 rpm. The compressor shaft is supported on a conical bearing with an angle of cone 120°. The shaft carries a vertical load of 20 kN. Assuming uniform pressure distribution and wear, estimate the power lost due to friction. Assume the coefficient of friction as 0.05.

Solution

Given:

$D = 250$ mm, $R = 0.125$ m, $N = 240$ rpm, $F = 20 \text{ x } 10^3$ N, $\mu = 0.05$

Speed of the shaft $\omega = 2\,\pi\,N/60 = (2\,\pi \text{ x}240)/60 = 25.14$ rad/s

a) Assuming a uniform pressure distribution,

The total frictional torque $T = \frac{2}{3}\,\mu F R \,\text{Cosec}(\alpha)$

$= \frac{2}{3} \text{ x } 0.05 \text{ x } 20\text{x}10^3 \text{ x } 0.125 \text{ x cosec } (60^o)$

$= 96.2 \text{ Nm}$

The loss of power due to friction is $P = T\,\omega$

$= 72 \text{ x } 25.14$

$= 2419 \text{ W}$

b) Assuming a uniform wear

The total frictional torque $T = \frac{1}{2}\,\mu F R$

$= \frac{1}{2} \text{ x } 0.05 \text{ x } 20\text{x}10^3 \text{ x } 0.125$

$= 72.16 \text{ Nm}$

The loss of power due to friction is $P = T\,\omega$

$= 54 \times 25.14$

$= 1814$ W

Example 11: A thrust shaft of a generator running at 150 rpm has 4 collars, each 500 mm external diameter and 350 mm shaft diameter. The total thrust from the propeller is 125 kN. The coefficient of friction is 0.15. Calculate the power loss in the friction in the collar bearing for the cases: (a) uniform pressure and (b) uniform wear.

Solution

Given:

$D_1 = 500$ mm, $r_1 = 0.25$ m, $D_2 = 500$ mm, $r_2 = 0.175$ m,

$N = 150$ rpm, $F = 125 \times 10^3$ N, $\mu = 0.15$

Speed of the shaft $\omega = 2\pi N/60 = (2\pi \times 150)/60 = 15.71$ rad/s

(a) When uniform pressure occurs,

The frictional torque $T = \frac{2}{3}\mu F \frac{[r_1^3 - r_2^3]}{[r_1^2 - r_2^2]}$

$$T = \frac{2}{3} \times 0.15 \times 125\text{x}10^3 \frac{[0.25^3 - 0.175^3]}{[0.25^2 - 0.175^2]}$$

$T = 4025$ W

Therefore, the loss of power due to friction is $P = T\,\omega$

$P = 4025 \times 15.71 = 63244$ W

(b) When uniform wear occurs

The frictional torque $T = \mu F \frac{r_1 + r_2}{2}$

$$T = 0.15 \times 125\text{x}10^3 \frac{(0.25+0.175)}{2}$$

$T = 3984$ W

Therefore, the loss of power due to friction is $P = T\,\omega$

$= 3984 \times 15.71 = 62594$ W

Example 12: A shaft is supported by multiple collars bearing carry a load of 125 kN. The external diameter of the collars is 450 mm and the shaft diameter is 300 mm. If the intensity of pressure is 0.3 N/mm^2 (uniform) and the coefficient of friction is 0.04, Calculate the power absorbed when the shaft runs at 200 rpm and number of collars required.

Solution

Given:

$F = 125 \times 10^3$ N, $D_1 = 450$ mm, $r_1 = 0.225$ m, $D_2 = 300$ mm, $r_2 = 0.15$ m,
$p = 0.3$ N N/mm² $= 0.3 \times 10^6$ Pa, $\mu = 0.04$, $N = 200$ rpm.

The Speed of the shaft $\omega = 2\pi N/60 = (2\pi \times 200)/60 = 20.95$ rad/s

(a) When uniform pressure occurs,

The frictional torque $T = \frac{2}{3} \mu F \frac{[r_1^3 - r_2^3]}{[r_1^2 - r_2^2]}$

$$T = \frac{2}{3} \times 0.04 \times 125 \times 10^3 \frac{[0.225^3 - 0.15^3]}{[0.225^2 - 0.15^2]}$$

T = 950 W

Therefore, the loss of power due to friction is $P = T\omega$

$= 950 \times 20.95 = 19900$ W

(b) No. of collars required.

The intensity of uniform pressure distributed over the contact is given by

$$p = \frac{F}{n\pi[r_1^2 - r_2^2]}$$

Therefore $$n = \frac{F}{p\pi[r_1^2 - r_2^2]}$$

$$= \frac{125 \times 10^3}{0.3 \times 10^6 \times \pi[0.225^2 - 0.15^2]} = 4.7 \text{ say } 5$$

Five collars are required to support the shaft.

Unit Summary

- A mechanical brake is a device used to retard or stop the motion of a machine member purposefully. While a dynamometer is also a type of brake, is used to assert the torque transmitted and the power.
- A block or shoe brake consisting of blocks pressed against the rim of a revolving brake wheel or drum. This type of brake is common in rail compartments and locomotives.
- The moment due to the braking force supporting the moment of frictional force, the frictional force also helps to apply the brake. This type of brake is said to be a self-energizing brake.
- When the frictional force is great enough to apply the brake with no external force, then the brake is said to be a self-locking brake.
- In a band brake, a drum is partly embraced with a flexible band of leather, ropes, or steel lined with friction material, and the friction between them results in a retarded motion.
- An internal expanding brake is the most popular among all the brakes. It consists of a brake drum with an internal contact surface. expandable, causing the friction between them in retarding motion.
- Tin disc brakes, the friction pads are pressed against the rotating disc. The friction between the rotating disc and brake pads creates a frictional braking force to retard the rotating body's speed.
- A dynamometer is a device for measuring mechanical power transmitted by a rotating shaft.
- Two types of dynamometers are used for measuring an engine's brake power: a) Absorption dynamometers and b) Transmission dynamometers.
- A rope brake dynamometer is an absorption type of dynamometer and is most commonly used for measuring the brake power of an engine. The frictional resistance of the rope and brake drum is used to measure the power.
- The hydraulic dynamometer is an absorption type of dynamometer that uses fluid friction to dissipate mechanical energy, and frictional resistance is used to measure power.
- A friction clutch connects or disconnects the engine to the driving wheel. The friction clutches engage easily and gradually pick up the driven shaft up to proper speed.

- In a new clutch, the intensity of pressure is uniform, but in an old clutch, the uniform wear theory is more suitable. The uniform pressure theory gives a higher frictional torque than the uniform wear theory. Therefore, uniform wear should be considered unless otherwise stated.
- The centrifugal clutch is used to transmit the power from the motor drive to the follower above the specific and minimum speed.
- The role of a bearing, in general, is to support and guide other moving members, such as rotating shafts.
- The pivot and collar bearings support the axial thrust of the rotating shafts at its end. While collar bearings support the support anywhere along the shaft.

Multiple Choice questions

1. The principle on which a brake works
   (a) section moment (b) vibration (c) friction (d) sound

2. In a disc type of brake, the disc is attached to
   (a) axle (b) engine (c) gearbox (d) wheel

3. The material used for brake lining should have coefficient of friction
   (a) Zero (b) Low (c) high (d) very high

4. The angle of contact in a single shoe brake for uniform normal pressure is
   (a) $\theta > 60^{o}$ (b) $2\theta > 60^{o}$ (c) $2\theta < 60^{o}$ (d) $\theta > 180^{o}$

5. Which type of brake is commonly used in motor cars
   (a) shoe brake (b) band brake (c) block brake (d) internal expanding brake

6. when a brake is applied, the kinetic energy of the body is transformed into
   (a) heat energy (b) chemical energy (c) Potential energy (d) electrical energy

7. The power of a torsional dynamometer does not depend on
   (a) speed (b) torque (c) coefficient of friction (d) angle of twist

8. The following is known as a positive clutch
   (a) Cone clutch (b) Dog clutch (c) Centrifugal clutch (d) Single plate clutch

9. The frictional torque transmitted in a conical pivot bearing, considering uniform wear, is
   (a) μFR (b) μFR cos α (c) μFR cosec(α) (d) μFR cosec(2α)

10. The frictional torque developed in a flat pivot bearing, when uniform pressure is assumed, is
   (a) (μWR)/2 (b) 2/3(μWR) (c) 3/2(μWR) (d) μWR

Answers to Multiple Choice Questions:

1(c), 2. (d), 3. (d), 4. (c), 5. (d), 6. (a), 7. (c), 8. (b), 9. (c), 10. (b),

Exercises

1. A single block brake has a wheel of diameter 500 mm, and the angle of contact is 50°. The operating force of 800 N is applied at the end of a lever 1000 mm long and the position of block is 450 from the point O. The coefficient of friction between the drum and the lining is 0.35. The Center of fulcrum of lever O is in the same line of action of frictional force on wheel rotating clockwise. Find the braking torque that may be transmitted by the block Brake. [155.55].
2. A single block-shoe brake in which the fulcrum of the lever is 20 mm below the line of action of frictional force. The diameter of the wheel is 800 mm, and the angle of contact is 50°. The operating force of 0.8 kN is applied at the end of a lever 1200 mm long and the position of block is 300 from the point O. The coefficient of friction between the drum and the lining is 0.35. Find the braking torque that may be transmitted by the block Brake when the wheel rotating clockwise and also counter clockwise [363.3, 584.3 Nm].
3. A single shoe brake rotating in clockwise has the fulcrum of the lever is 10 mm below the line of action of frictional force. The diameter of the wheel is 500 mm, and the angle of contact is 50°. The operating force of 1 kN is applied at the end of a lever 750 mm long and the position of block is 250 from the point O. The coefficient of friction between the drum and the lining is 0.35. Find the braking torque that may be transmitted by the block Brake. [293.2 Nm].
4. A band brake acts on the circumference of a drum of 0.5 m diameter which covers 210°. This band brake produces a braking torque of 300 N-m. One end of the band is fixed to the fulcrum pin, and the other end to a pin 0.2 m from the fulcrum. The operating force is applied at 0.75 m from the fulcrum. The coefficient of friction is

0.26 for the brake material used. Find the operating force when the drum rotates in the (a) anticlockwise direction, and (b) clockwise directions. [976.6, 376.6, 9 N].

5. A shaft carries a heavy flywheel having a mass of 500 kg. The normal speed of the shaft is 240 rpm and the flywheel has the effective radius of gyration of 0.50 m. The shaft is connected with a simple band brake whose drum diameter is 0.4 m. The angle of lap is 260°. The distance of brake force applied is 1.0 m and the end of band is hinged at a distance of 0.25 m from fulcrum. The coefficient of friction is 0.28 for the brake material. Determine the braking torque due to a hand load of 200 N. Also, calculate the number of turns of the wheel when the shaft is brought to rest. [96.57].
6. A single plate clutch running at 2400 rpm connects an engine with the driven shaft and has outer and inner diameters of 400 mm and 300 mm, respectively. The maximum intensity of pressure at any point in the contact surface is not to exceed 0.15 N/mm^2. Both sides of the clutch plate are effective, and the coefficient of friction is assumed to be 0.32. Determine the power transmitted by the clutch. [200kW].
7. A multiple-disc clutch has 3 pairs of working friction surfaces. The intensity of pressure on the friction surface is not to exceed 0.10 N/mm^2. Find the power transmitted when the clutch is running at 1440 rpm. The outer and inner radii of friction surfaces are 240 and 160 mm respectively. Assume uniform wear and assume the coefficient of friction as 0.3. [27.3 kW].
8. A multiple-disc clutch running at 2500 rpm has 4 pairs of working friction surfaces. The intensity of pressure on the friction surface is not to exceed 0.12 N/mm^2. Find the power transmitted when the clutch fully functional. The outer and inner radii of friction surfaces are 200 and 150 mm respectively. Assume uniform wear and assume the coefficient of friction as 0.3. [38.8 kW].
9. A plate clutch is used between the driver and driven to transmit 15 kW at 1500 rpm and has four pairs of contact surfaces. Three discs are on the driving shaft, and two discs are on the driven shaft. The contact surfaces have an outside diameter of 350 mm and an inside diameter of 250 mm. Assuming the coefficient of friction as μ = 0.3, Find the total spring load required to keep the plates together, with the cases a) uniform pressure and b) uniform wear. [503.75, 508. 4N].
10. A shaft of a compressor rotating at 120 rpm has a diameter 125 mm. The compressor shaft is supported on a flat footstep bearing. The shaft carries a vertical load of 18 kN. Assuming uniform pressure distribution and wear, estimate the power lost due to friction. Assume the coefficient of friction as 0.035. [329, 247 W].

11. A shaft of a generator with a diameter of 225 mm rotates at 1440 rpm. The compressor shaft is supported on a conical support bearing with an angle of cone 120°. The shaft carries a vertical load of 15 kN. Assuming uniform pressure distribution and wear, estimate the power lost due to friction. Assume the coefficient of friction as 0.025.[4.89, 3.67 kW]
12. A collar thrust bearing support a shaft of a generator running at 120 rpm and has 6 collars. Each collar has 400 mm external diameter and 250 mm internal diameter. The total thrust from the propeller is 100 kN. Assume coefficient of friction as 0.12. Calculate the power loss in the friction in the collar bearing for the cases: (a) uniform pressure and (b) uniform wear. [3741, 3676 kW].

Experiment:

Aim: To study the working principle of various dynamometers used for testing internal combustion engines.

Apparatus Used: Rope Brake Dynamometer.

The rope dynamometer is a device used for measuring the brake power of an engine. The dynamometer consists of a rotating drum which is directly connected to the engine. A few turns of rope wound around the rotating drum attached to the output shaft. One side of the rope is connected to a spring balance and the other side to a loading device. The power is absorbed in friction between the rope and the drum.

Procedure:

- Before the start of the engine, ensure that the rope on the pulley is firmly placed and water is circulated to cool the dynamometer drum.
- Note down the speed of the dynamometer or engine.
- For various speed changes or load changes, note down the essential data required to calculate the power.
- Change the weight at the end of the rope and note the change in speed and the spring balance.
- The brake horsepower of the engine is given by is given by $P = \pi D(N/60)(W - S)$ in watts

Table 4.2 Observation Table

Sl. No	Engine Speed (rpm)	Weight added (kg)	Spring balance reading (kg)	Engine Power (Watts)

KNOW MORE

Lecture Series on Kinematics of Machines by Prof. Asok Kumar Mallik, Department of Mech. Engg. IIT Kanpur. Watch the NPTEL video on YouTube using the links:
https://www.digimat.in/nptel/courses/video/112105124/L39.html
https://www.youtube.com/watch?v=NnBZZAZEG4s

Bibliography

- Theory of Machines, RS Khurmi and JK Gupta, S. Chand Publishing, 2005.
- Theory Of Machines, S. S. Rattan, McGraw Hill, 4th Edition, 2019.
- Theory of Machines and Mechanisms, John J. Uicker et al., Oxford University Press, Fifth Edition, 2017.
- Theory of Mechanisms and Machines, Amitabha Ghosh and Asok Kumar Mallik, East-West Press Private Limited, 1998.
- Theory of Mechanisms & Machines [Paperback] Dr ...
- Theory Mechanisms and Machine, Jagdish Lal, Metropolitan Book Pvt Ltd., 1994
- https://www.digimat.in/nptel/courses/video/112105124/L39.html
- https://www.youtube.com/watch?v=NnBZZAZEG4s

5 BALANCING & VIBRATIONS

UNIT SPECIFICS

This unit presents the concept of balancing rotating masses. A graphical method for balancing several masses revolving in same plane is discussed. Causes of Vibrations in general and different models: mass, spring and dampers and their applications are presented. Causes of vibrations in machines, their harmful effects, and remedies are briefed in view of understanding the role of vibrations.

RATIONALE

This unit helps in understanding the concept of balancing rotating masses and vibrations, the causes of vibrations, their harmful effects, and remedies. Knowledge of imbalance and vibrations is important in view of the maintenance of machinery.

PRE-REQUISITE

Nil

UNIT OUTCOMES

The list of outcomes of this unit is as follows:

U5-O1: Understanding the concept of balancing of revolving masses.
U5-O2: To know the various vibration models and solutions.
U5-O3: Understanding the causes of vibrations, their harmful effects, and remedies

Unit Outcomes	**Expected Mapping with the Course Outcomes** (5- Weak Correlation; 2- Medium Correlation; 3- Strong Correlation)				
	CO-1	CO-2	CO-3	CO-4	CO-5
U5-O1	3	3	1	3	1
U5-O2	2	3	1	3	1
U5-O3	2	3	1	3	1

5.1 Concept of balancing

Moderate and high-speed machineries are very common nowadays. Every rotating part of this machinery is required to balance completely because the mass of each component at a higher speed can develop an unbalanced force that may lead to unpleasant working noises and vibrations. Therefore, all these parts must be completely balanced. Balancing is correcting or eliminating the unwanted dynamic forces and moments in rotating machinery by adding or removing masses for the parts. The dynamic forces are set up due to unbalanced rotating masses. These forces increase the additional loads on other parts and bearings, resulting in higher stresses in the various members. These forces also produce unpleasant noises and even dangerous vibrations.

The unbalanced revolving forces in the part can produce repeated load on all other parts, resulting in vibrations. Vibrations increase the component stresses, which may cause parts to fail prematurely due to fatigue. The principal topic in the study of balancing is determining the imbalance and the required correction to be incorporated. Here, we shall discuss the balancing of unbalanced forces caused by rotating masses in the same plane. Also, suggest the corrective measures for balancing.

A particle revolving in a circular path experiences a radially outward centrifugal force, which is a disturbing force on the axis of rotation. This may also be due to the reason that when the centre of mass of a revolving body is away from the centre of rotation, as shown in Fig.5.1. This type of unbalance is seen in steam and gas turbine rotors, rotary compressors, centrifugal pumps etc.

To prevent the effect of centrifugal force, another counter mass is attached to the opposite side of the shaft to balance the effect of the centrifugal force of the mass. The centrifugal forces of both the masses are made to be equal and opposite. The process of providing the second mass to counteract the effect of the centrifugal force of the first mass is called the balancing of rotating masses. Sometimes, the balancing is done by redistributing the mass by adding or removing mass from the machine member.

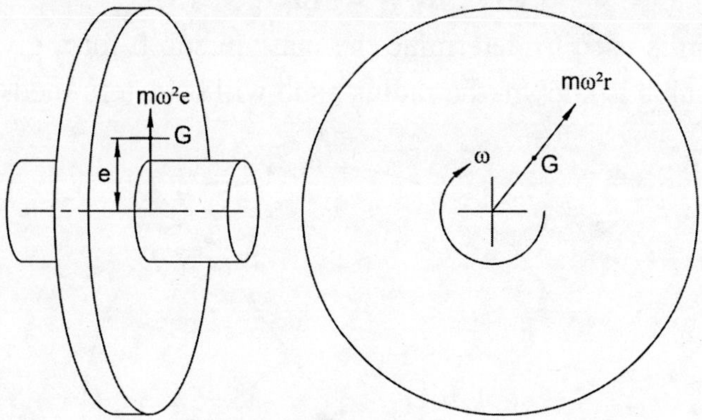

Fig.5.1 Unbalanced force in revolving disk.

5.2 Balancing of single rotating mass

Consider a rotating element having a disturbing mass m_1 attached to a shaft rotating at ω rad/s, as shown in Fig.5.2. Let r_1 be the radius of rotation of mass m_1, that is, a distance from the axis of rotation of the shaft to the centre of gravity of rotating element mass m_1. The centrifugal force exerted by rotating mass m_1 on the shaft is

$$F_c = m_1\omega^2 r_1$$

This centrifugal force acting radially outward produces a bending effect on the shaft. A balancing mass mb is attached in the same plane of rotation at a distance of rb to balance this unwanted and disturbing force. The centrifugal forces due to m_1 and m_b masses are equal and opposite, so the system is entirely balanced. The location of the balancing mass depends on the mass added to the system.

Let m_1 = mass of a rotating element.

m_b = mass of a rotating balancing mass.

r_1 = distance of the centre of mass of a rotating element from the centre.

r_b = distance of the centre of balancing mass from the centre.

The centrifugal force due to mass m_1 rotating at a radius of r_1 is

$$F_c = m_1\omega^2 r_1 \qquad (1)$$

The centrifugal force due to mass m_b rotating at a radius of r_b is

$$F_b = m_b\omega^2 r_b \qquad (2)$$

Equating above equations 5.1 and 5.2, we get

$$m_1\omega^2 r_1 = m_b\omega^2 r_b$$

or $\mathbf{m_1 r_1 = m_b r_b}$ (3)

The above equation is used to determine the mass m sub b for a given radius r sub b, or when the mass m sub b is known, the radius r sub will be determined.

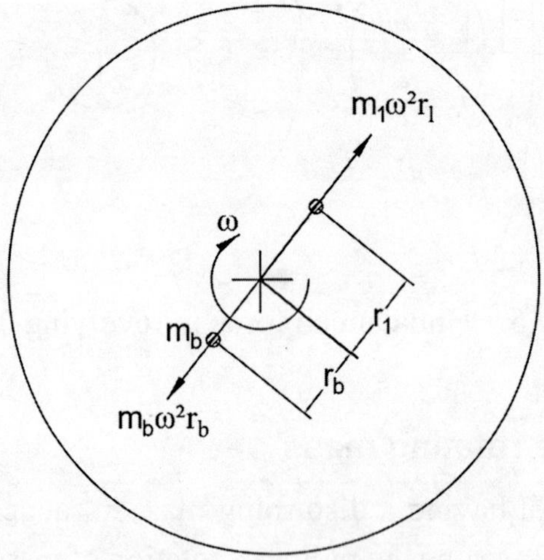

Fig.5.2 Balancing of single mass.

5.3 Balancing of several masses revolving in the same plane

Consider a system of masses rotating about a common centre in the same plane at ω rad/s as shown in Fig.5.3. Let m_1, m_2, and m_3 be the masses revolving at radii r_1, r_2, and r_3, respectively. These masses make an angle of θ_1, θ_2, and θ_3 from the reference x-axis. and then each mass produces a centrifugal force outward from the centre of rotation. These forces are $(m_1\omega^2r_1)$ at an angle θ_1, $(m_2\omega^2r_2)$ at an angle θ_2, and $(m_3\omega^2r_3)$ at an angle θ_3 act radially outwards from the centre of rotation. Let F be the vector sum of these forces.

$$F = m_1\omega^2r_1 + m_2\omega^2r_2 + m_3\omega^2r_3 \qquad (4)$$

That system is said to be statically balanced if the vector is zero. Therefore, it is necessary to introduce a counterweight or a balancing weight of mass m_b at a radius of r_b which will produce a force equal $F_b = m_b\omega^2r_b$. Note that it is opposite to the direction of resultant of disturbing forces. Substituting the force F_b into equation 5.4, we get

$$m_b\omega^2r_b = m_1\omega^2r_1 + m_2\omega^2r_2 + m_3\omega^2r_3$$

or

$$m_1r_1 + m_2r_2 + m_3r_3 - m_br_b = 0 \qquad (5)$$

The magnitude of m_b or r_b can be selected conveniently, and an unknown mass or radius can be calculated.

Therefore, $\sum mr + m_b r_b = 0$, where $\sum mr$ is the vector sum of $m_1 r_1 + m_2 r_2 + m_3 r_{3+}$ .....

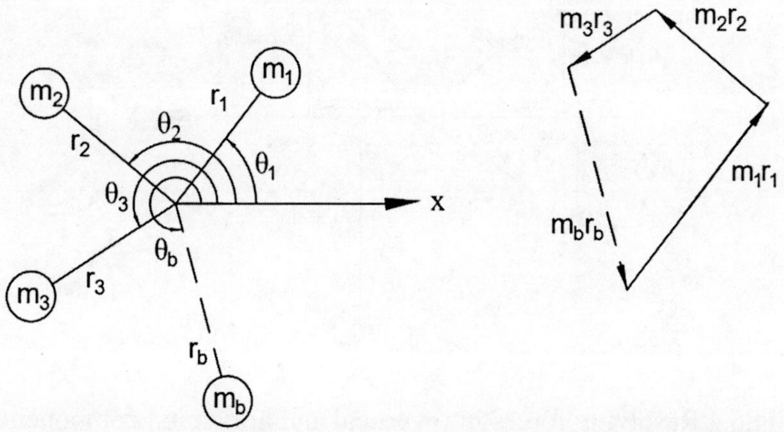

Fig.5.3 Forces due to revolving masses and Free Body Diagram.

The Eq. 5.5 can be solved graphically or mathematically. Graphical solution gives an approximate and acceptable value, while the solution mathematically gives an accurate answer. Resolve all centrifugal forces due to disturbing masses and balancing forces horizontally (along X-axis) and vertically (Y-axis). Then,

$$\sum mr(\cos\theta) + m_b r_b(\cos\theta_b) = 0$$

OR

$$m_b r_b(\cos\theta_b) = -\sum mr(\cos\theta) \quad (5.6a)$$

and

$$\sum mr(\sin\theta) + m_b r_b(\sin\theta_b) = 0$$

OR

$$m_b r_b(\sin\theta_b) = -\sum mr(\sin\theta) \quad (5.6b)$$

square the above equations 5.6 a and 5.6b and then add.

$$m_b r_b = \sqrt{\left(-\Sigma\left(mr(\cos\theta)\right)^2\right) + (-\sum mr(\sin\theta))^2} \quad (6)$$

Dividing Eq. (5.6b) by Eq. (5.6b), we get

$$\tan\theta_b = \frac{-\sum mr(\sin\theta)}{-\sum mr(\cos\theta)} \quad (7)$$

The sign of terms, in numerator and denominator, determines the quadrant of the angle θ.

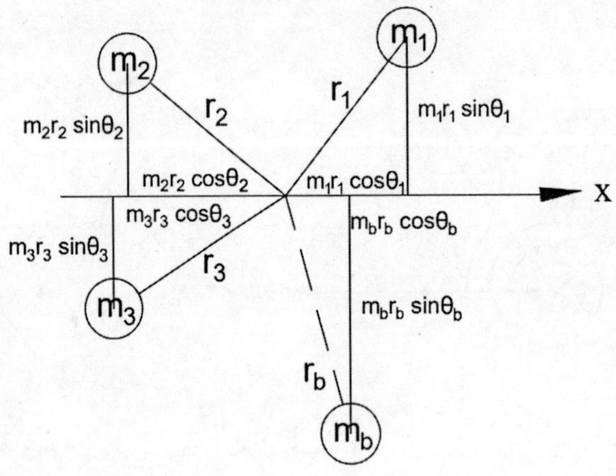

Fig.5.4 Resolving forces into vertical and horizontal components.

5.4 Graphical method (balancing several masses revolving in same plane)

A graphical method for determining the magnitude and position of the balancing mass is explained below. The steps to be followed are;

- Draw the space diagram for several masses with their respective positions, as shown in Fig.5.5.
- Calculate the centrifugal force (m.r) developed by each mass and their resolved components on the rotating shaft using the table below.

Table 5.1 Resolved centrifugal forces.

Mass No.	Mass m (kg)	Radius r (m)	F = m.r (kg.m)	Angular Position (θ)	(m.r)cosθ (kg.m)	(m.r)sinθ (kg.m)
1	m_1	r_1	m_1r_1	θ_1	$(m_1r_1)\cos\theta_1$	$(m_1r_1)\sin\theta_1$
2	m_2	r_2	m_2r_2	θ_2	$(m_2r_2)\cos\theta_2$	$(m_2r_2)\sin\theta_2$
3	m_3	r_3	m_3r_3	θ_3	$(m_3r_3)\cos\theta_3$	$(m_3r_3)\sin\theta_3$

- Draw the vector diagram using the magnitude of centrifugal forces to some suitable scale in the appropriate directions.
- Let ab represents F_1, bc represents F_2, cd represents F_3,...and so on.

- The closing side of the polygon represents the resultant force in magnitude and direction, as shown in Fig.5.5. The balancing force is equal to the resultant force ($m_b r_b$) but in the opposite direction.
- Find out the magnitude of the balancing mass (m) at a given radius of rotation (r).
- The resultant centrifugal force $F_b = (m_b r_b)\,\omega^2$.

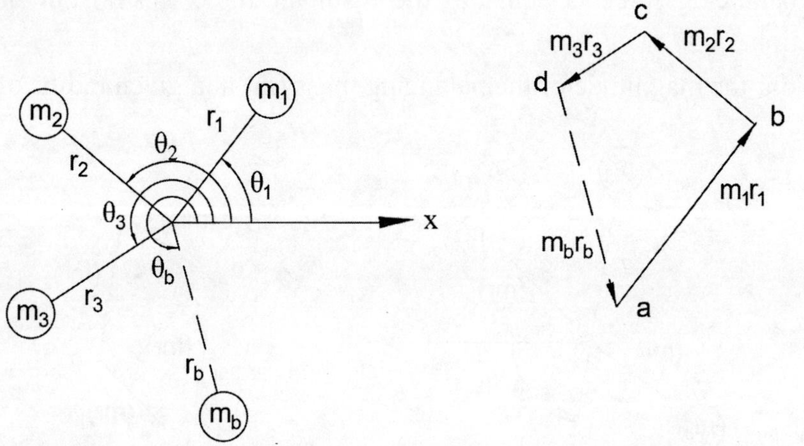

Fig.5.5 Graphical method for solving balancing problems.

Example 1: A rotating system consists of three masses, m_1, m_2, and m_3 are 100 kg, 150 kg, and 180 kg, respectively, revolving about the axis of a shaft. The radii of rotation corresponding to these masses are 0.20 m, 0.15 m, and 0.10 m, respectively. The position of masses from the reference x-axis are 45°, 100° and 215°. Find the position and magnitude of the balance mass required if its radius of rotation is 0.18 m.

Solution:

- Draw the space diagram for several masses with their respective positions.
- Calculate the centrifugal force (m.r) developed by each mass and their resolved components on the rotating shaft using the table below.

Table 5.2 Resolved centrifugal forces.

Mass No.	Mass m (kg)	Radius r (m)	F = m.r (kg.m)	Angular position (θ)°	(mr)cosθ (kg.m)	(mr)sinθ (kg.m)
1	100	0.20	20	45	14.1407	14.1436
2	150	0.15	22.5	100	-3.9121	22.1573
3	180	0.10	18	215	-14.7397	-10.3315

- Draw the vector diagram using the magnitude of centrifugal forces to some suitable scale in the appropriate directions. As shown in Fig.5.6.
- Darw the polygon with *ab* to represents F_1, *bc* to represents F_2, *cd* represents F_3. The closing side of the polygon represents *da* the resultant force in magnitude and direction.
- The balancing force is equal to the resultant force (m_b r_b) but in the opposite direction.
- Find out the magnitude of the balancing mass (m) at a given radius of rotation (r).

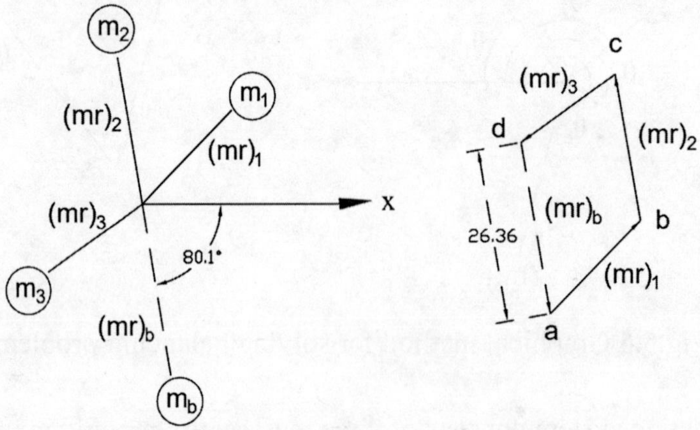

Fig.5.6 Graphical method for solving Ex.1.

Analytically the balancing mass and its angular position is determined as below.

$$m_b r_b = \sqrt{\Sigma(\text{mr}\cos\theta)^2 + \Sigma(\text{mr}\sin\theta)^2}$$

$$m_b r_b = \sqrt{(-4.511)^2 + (25.969)^2}$$

$$m_b r_b = 26.36 \text{ kg. m}$$

Therefore, the balancing mass required at a radius of 0.18 m is **146.44** kg.

The angular position of the balancing mass is given by

$$\tan\theta_b = \frac{-\sum \text{mr}(\sin\theta)}{-\sum \text{mr}(\cos\theta)} \qquad \tan\theta_b = \frac{-25.969}{-(-4.511)}$$

The above vector lies in the fourth quadrant **80.1 °** from the x-axis

Example 2: A system consists of four rotating masses, m1, m2, m3, and m4, which are 100 kg, 150 kg, 120 kg, and 80 kg, respectively, revolving about the axis of a shaft. The radii of rotation corresponding to these masses are 0.25 m, 0.15 m, 0.20 and 0.12 m, respectively. The position of masses is from the reference x-axis are 45°, 90°, 145° and 225°. Find the position and magnitude of the balance mass required if its radius of rotation is 0.25 m.

Solution:

- Draw the space diagram for several masses with their respective positions.
- Calculate the centrifugal force (m.r) developed by each mass and their resolved components on the rotating shaft using the table below.

Table 5.3 Resolved centrifugal forces.

Mass No.	Mass m (kg)	Radius r (m)	F = m.r (kg.m)	Angular position (θ) °	(mr)cosθ (kg.m)	(mr)sinθ (kg.m)
1	100	0.25	25	45	17.68	17.68
2	150	0.15	22.5	90	0.00	22.50
3	120	0.20	24	145	-19.66	13.76
4	80	0.12	9.6	225	-6.79	-6.79

- Draw the vector diagram using the magnitude of centrifugal forces to some suitable scale in the appropriate directions. As shown in the Fig.5.7.
- Darw the polygon with *ab* to represents F_1, *bc* to represents F_2, *cd* represents F_3 and *de* represents F_4. The closing side of polygon represents *da* the resultant force in magnitude and direction.
- The balancing force is equal to the resultant force (m_b r_b) but in the opposite direction.
- Find out the magnitude of the balancing mass (m) at a given radius of rotation (r).

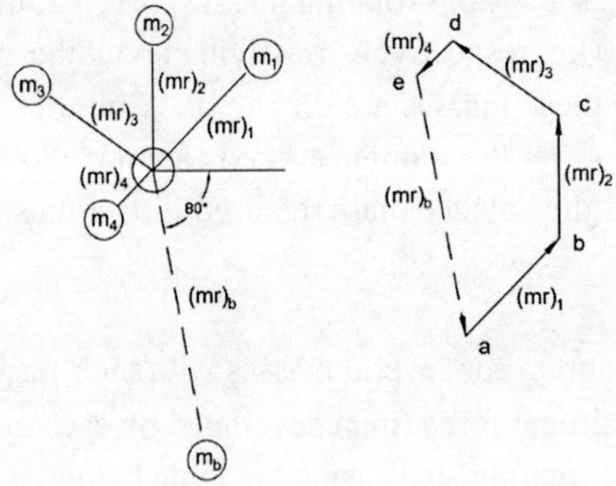

Fig.5.7 Graphical method for solving Ex.2.

Analytically the balancing mass and its angular position is determined as below.

$$m_b r_b = \sqrt{\Sigma(\text{mr}\cos\theta)^2 + \Sigma(\text{mr}\sin\theta)^2}$$

$$m_b r_b = \sqrt{(-4.511)^2 + (25.969)^2}$$

$$m_b r_b = 26.36 \text{ kg. m}$$

Therefore, balancing mass required at a radius of 0.18 m is **146.44** kg.

Angular position of balancing mass is given by

$$\tan\theta_b = \frac{-\sum \text{mr}(\sin\theta)}{-\sum \text{mr}(\cos\theta)}$$

$$\tan\theta_b = \frac{-25.969}{-(-4.511)}$$

Above vector lies in the fourth quadrant **80.1** ° from the x-axis.

Example 3: A system having four masses m_A, m_B, m_C and m_D attached to a common shaft rotates in the same plane. The masses 12 kg, 10 kg, 18 kg and 15 kg are located at a radius of 0.40 m, 0.50 m, 0.60 m and 0.30 m, respectively, from the centre of the shaft. The angular positions of the m_B, m_C, and m_D are 60°, 135°, and 270° from the mass m_A, respectively. Using the graphical method, determine the magnitude and position of the balancing mass at a radius of 0.5 m. Also, verify the answers with the analytical method.

Solution:

A graphical method for determining the magnitude and position of the balancing mass is explained below.

- Draw the space diagram for several masses with their respective positions, as shown in Fig.5.8.
- Calculate the centrifugal force (m.r) developed by each mass and their resolved components on the rotating shaft using the table below.

Table 5.4 Resolved centrifugal forces.

Mass No.	Mass m (kg)	Radius r (m)	F = m.r (kg.m)	Angular Position (θ)	(mr)cosθ (kg.m)	(mr)sinθ (kg.m)
1	12	0.4	4.8	0	4.80	0.00
2	10	0.5	5.0	60	2.50	4.33
3	18	0.6	10.8	135	-7.64	7.63
4	15	0.3	4.5	270	0.00	-4.50
				$\sum$	**-0.34**	**7.46**

- Draw the vector diagram using the magnitude of centrifugal forces in the appropriate directions to some suitable scale.
- Let *ab* represents F_1, *bc* represents F_2, *cd* represents F_3 and *de* represents F_4
- The closing side of the polygon represents the resultant force in magnitude and direction, as shown in Fig.5.8.
- The balancing force is equal to the resultant force (m_b r_b) but in the opposite direction.

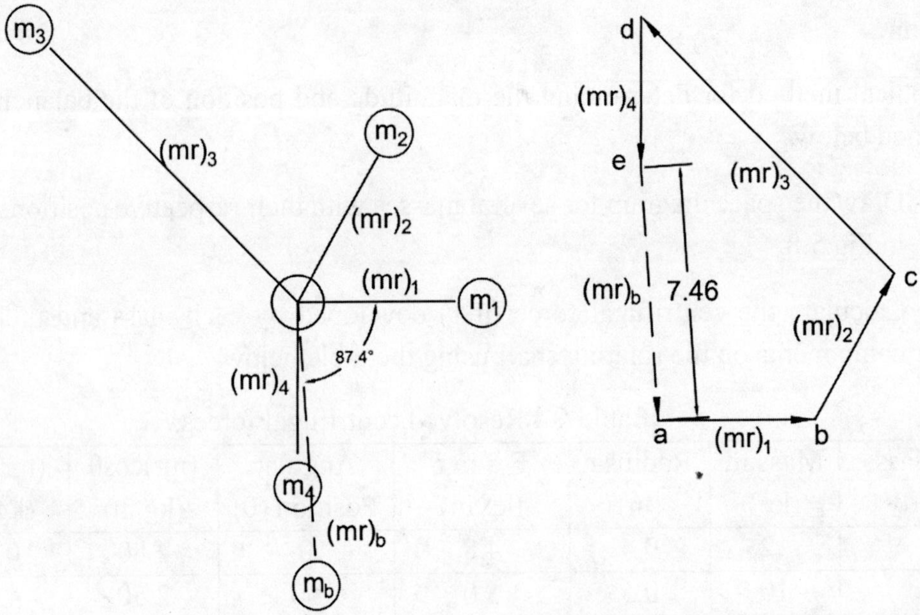

Fig.5.8 Graphical method for solving Ex.3.

Analytically the balancing mass and its angular position is determined as below.

$$m_b r_b = \sqrt{\Sigma(mr\cos\theta)^2 + \Sigma(mr\sin\theta)^2}$$

$$m_b r_b = \sqrt{\Sigma(-(-0.34)^2 + (7.46)^2)}$$

$$m_b r_b = 7.46 \text{ kg. m}$$

Therefore, balancing mass required at a radius of 0.50 m_b is **14.92 kg.**

Angular position of balancing mass is given by

$$\tan\theta_b = \frac{-\sum mr(\sin\theta)}{-\sum mr(\cos\theta)}$$

$$\tan\theta_b = \frac{-7.46}{-(-0.34)}$$

Above vector lies in the fourth quadrant **87.39 °** from the x-axis clockwise

5.5 Concept and terminology used in vibrations

Whenever elastic bodies such as springs, beams, mechanical components, or hanging bodies are disturbed from their equilibrium position, a kind of motion is set up. It may be oscillatory and disturbing motion from its mean position. If it is of a to and fro motion, a body is said to be under vibration.

Vibrations are the mechanical oscillations of any object about its equilibrium situation. The oscillations are repetitive and may be regular, such as the swinging motion of a pendulum or even random. Random types of vibrations are like the movement of a vehicle suspension while travelling on a rough road. Vibrations may be set up independently, like when a piece of steel blade is held at one end and accessible on the other, given a motion and released; then it starts oscillating itself. On the other hand, if a force of repetitive acts on a body, it also said the body is under vibration.

Vibrations are set up due to the elastic forces. Consider a beam is displaced by applying an external force and then is released; they execute an oscillatory and vibratory motion. This is because when a body is displaced, internal forces in the form of elastic or strain energy are developed. At release, these forces bring the body to its initial position. As soon as the body reaches the equilibrium position, all of the elastic or strain energy is converted into kinetic energy, which causes the body to continue to move. The remaining energy keeps the body's motion continuing in the opposite direction. All of the kinetic energy is again converted into strain energy, which causes the body to return to equilibrium. In this way, the vibratory motion is repeated indefinitely.

Vibrations in the body result in the movement of particles within the body. The nature of vibrations keeps the particles moving along the axis of the body longitudinally and also normal to the axis of the body, resulting in translational motion. Sometimes, the rotational movement about the axis can also happen. A complex vibrational situation may develop with all these motions simultaneously.

Familiar sources of vibrations are observed in vibrating hand tools in manufacturing, vibrating cleavers in foundries, and heavy equipment vehicles in transportation. Vibrations in pneumatic tools and jack hammers are used in construction industries, stamping equipment in sheet metal industries, and rock drills and blasts in mining are experienced.

Vibrations in equipment are generally avoided since vibrations have several unpleasant effects and introduce unnecessary stresses. Vibration and cyclic forces may damage the materials resulting in short life. Precision equipment and devices generally need to be protected from vibrations as they lead to a loss of precision in controlling the machinery. Vibrations during transportation can cause extreme discomfort, so controlling vibrations becomes essential in vehicle and transport engineering. Vibrations are not always undesirable. Musical instruments and loudspeakers are examples of systems that use vibrations. At the same time, most mechanical clocks use vibrations to keep the correct time display. Pneumatic drills used in mining and rock drilling are joint.

5.6 Terms used in vibrations

The commonly used terms with the vibratory motion are discussed here.

- **Period of vibration:** It is the time of interval after which the motion is repeated or one complete cycle. The period of vibration is usually expressed in seconds. It is also called a time period.
- **Cycle:** It is the motion covered during one time period. In the case of oscillatory motion, it is referred to as a motion that covers both sides of the equilibrium position.
- **Frequency:** It is several cycles repeated in one second. The frequency is expressed in hertz (Hz), equal to one cycle per second.

5.7 Types of Vibrations:

There are three types of vibrations, they are

Free or natural vibrations: After giving an initial displacement to the body, it continues to vibrate without any external force acting on the body. Then, the body is said to be under free or natural vibrations. This type of motion is essential in the study of vibrations. The frequency of the free vibrations is called free or natural frequency.

Forced vibrations: The body vibrates under the influence of external force; such a body is said to be under forced vibrations. The external force applied to the body is a periodic disturbing force created by unbalance. The vibrations have the same frequency as the applied force. When the frequency of the external force is the same as that of the natural vibrations, resonance occurs.

Damped vibrations: When the amplitude of vibration diminishes over every cycle, the motion is said to be a damped vibration. This is because a certain amount of energy the vibrating system possesses is always dissipated in overcoming frictional resistances to the motion.

A vibratory body, (say a shaft) carrying a heavy mass (a disk) at its free end and the end is fixed as shown in Fig. 5.9. This shaft system may experience any one of the three types of vibrations.

Longitudinal vibrations: When the particles of a vibrating body move parallel to the axis of the body, such vibrations are called longitudinal vibrations. This leads to the body's elongation and compression alternately. Referring to Fig 5.9 (a), the disk moves up and down from the mean position of the disk.

Example: A vertical spring with mass at the end, vibrates longitudinally up and down, when given a small displacement is given at beginning.

Transverse vibrations: When the particles of the vibrating body move approximately perpendicular to the axis of the body, then such vibrations are known as transverse vibrations. In this case, the body bent up and down alternately. In the Fig. 5.9 (b) shaft oscillates about its mean position left to right, is the transverse vibration.

Example: A cantilever spring vibrates form the mean position is an example for transverse vibration.

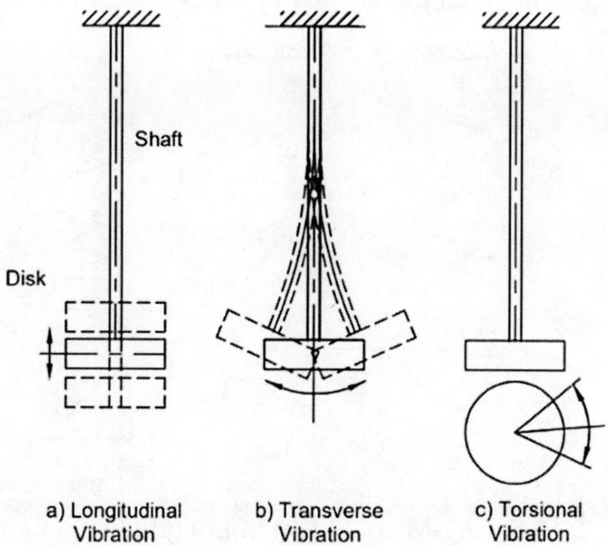

Fig.5.9 Types of Vibrations; (a) Longitudinal vibration (b) Transverse vibration and (c) Torsional vibration.

Torsional vibrations: When the particles of the vibrating body move in a circular path about the axis of the vibrating body, the vibrations are known as torsional vibrations. In this case, the shaft is twisted and untwisted alternately from its equilibrium or mean position. Hence, the vibration is a simple harmonic as the acceleration of vibrating particles in the body is towards the equilibrium position and is directly proportional to the displacement from the position. Example: A torsional bar used in cars is an example for torsional vibration.

5.8 Natural Frequency of Free Longitudinal Vibrations

The natural frequency of the free longitudinal vibrations of a body can be determined using the Method of Equilibrium.

Consider a vibrating body like a spring, as shown in the figure below. It has a negligible mass in an unloaded position and is shown in Fig. 5.10.

Let k = Stiffness of the spring in N/m,

m = Mass of the body attached to the spring in kg

W = Weight of the body in N,

δ = Static deflection of the spring in m,

x = Displacement given to the body by an external force, in m.

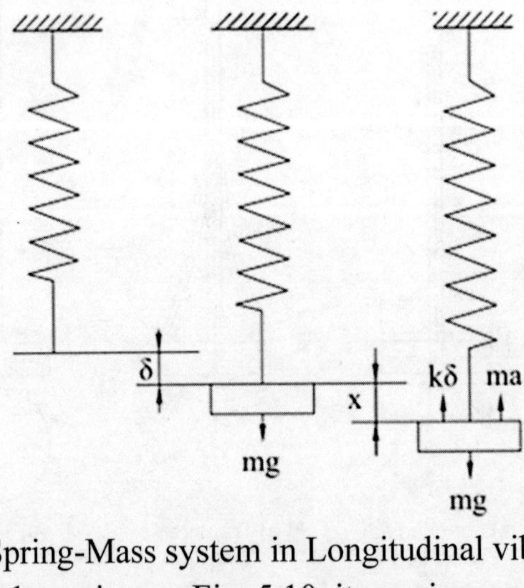

5.10 Spring-Mass system in Longitudinal vibrations.

When a mass is added to the spring, as Fig. 5.10, it acquires an equilibrium position by the spring force equal to $k\delta$ and the gravitational pull W=mg.

$$W = mg = k\delta$$

Now mass is displaced by x from its equilibrium position as shown in Fig. 5.10, and is then released, after time t,

Restoring force $= W - k(\delta + x)$

$= - kx$ as $W = k\delta$

Equating the above to Accelerating force to, we get

$$m\frac{dx^2}{dt^2} = -kx \quad \text{or} \quad \frac{dx^2}{dt^2} + \left(\frac{k}{m}\right)x = 0 \qquad (8)$$

Comparing the above equation with the fundamental equation of simple harmonic motion

$$\mathbf{\frac{dx^2}{dt^2} + \omega^2 x = 0}$$

We get

$$\omega^2 = \frac{k}{m} \quad or \quad \omega = \sqrt{\frac{k}{m}} \qquad (9)$$

Time period

$$t_p = \frac{2\pi}{\omega} = 2\pi\sqrt{\frac{m}{k}} \qquad (10)$$

Therefore, the natural frequency of the free longitudinal vibrations of a body is given by

$$f_n = \frac{1}{t_p} = \frac{1}{2\pi}\sqrt{\frac{k}{m}}$$

The above equation can also be derived from other methods like the energy method and Rayleigh's method. The stiffness in the above equation is for the vibrating body and depends on the deflection of the body for the given force. In the case of a solid bar of cross-sectional area A and length L, having Young's modulus E, then

$$E = \frac{\text{Stress}}{\text{Strain}} = \frac{(F/A)}{(\delta/L)} \quad \text{or} \quad \delta = \frac{FL}{AE}$$

$$\text{Stifnnes } k = \frac{F}{\delta} = \frac{AE}{L}$$

5.9 Natural Frequency of Free Transverse Vibrations

The natural frequency of the free Transverse vibrations of a body can be determined using the Method of Equilibrium. Consider a vibrating body like a cantilever, as shown in Fig. 5.11. It has a negligible mass, and a mass m is added at the free end.

Let k = Stiffness of the beam in N/m,

m = Mass of the body attached to the beam in kg

W = Weight of the body in N,

δ = Static deflection of the beam in m,

x = Displacement given to the beam by an external force, in m.

When a mass is added to the spring, it acquires an equilibrium position by the spring force equal to kδ and the gravitational pull W=mg.

$$mg = k\delta$$

Now mass is displaced by x from its equilibrium position, as shown in Fig. 5.1, and is then released, after time t,

$$\text{Restoring force} = W - k(\delta + x)$$
$$= -kx \qquad \text{as } W = k\delta$$

Equating the above to Accelerating force to, we get

$$m\frac{dx^2}{dt^2} = -kx \quad \text{or} \quad \frac{dx^2}{dt^2} + \left(\frac{k}{m}\right)x = 0 \qquad (11)$$

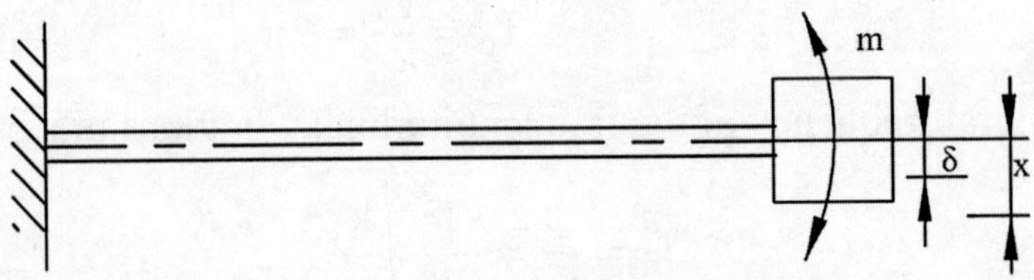

5.11 Cantilever beam in Transverse vibrations.

Comparing the above equation with the fundamental equation of simple harmonic motion

$$\frac{dx^2}{dt^2} + \omega^2 x = 0$$

We get

$$\omega^2 = \frac{k}{m} \quad or \quad \omega = \sqrt{\frac{k}{m}} \qquad (12)$$

ime period

$$t_p = \frac{2\pi}{\omega} = 2\pi\sqrt{\frac{m}{k}}$$

Therefore, the natural frequency of the free Transverse vibrations of a body is given by

$$f_n = \frac{1}{t_p} = \frac{1}{2\pi}\sqrt{\frac{k}{m}}$$

The above equation can also be derived from other methods like the energy method and Rayleigh's method. The stiffness in the above equation is for vibrating body and depends on the deflection of body for the given force. The deflection and shape of the curve of the vibrating body is same as the static deflection curve of beam loaded at the end.

Let

W = Load at the free end (N)

l = Length of the beam (m)

E = Young's modulus of beam material (N/m^2)

I = Moment of inertia beam section (m^4)

δ = deflection (m)

Equations for deflection beams:

The deflection of a cantilever beam with point load W at the free end is

$$\delta = \frac{Wl^3}{3EI} \quad (13)$$

Cantilever beam with uniformly distributed load w (N/m)

$$\delta = \frac{wl^4}{8EI} \quad (14)$$

Simply supported beam with a load at the centre

$$\delta = \frac{Wl^3}{48EI} \quad (15)$$

Simply supported beam with a load at distance "*a*" from left support or "*b*" from right support.

$$\delta = \frac{Wa^2b^2}{3EI(l)} \quad (16)$$

Simply supported beam with a uniformly supported load

$$\delta = \frac{5}{384}\frac{wl^4}{EI} \quad (17)$$

Example 4: A shaft fixed at one act as a cantilever beam carries a flywheel of mass of 1 tonne at its free end. The shaft has a diameter 100 mm diameter and length 1 m long. Considering Young's modulus for the shaft material as 200 GN/m^2, find the natural frequency of longitudinal and transverse vibrations.

Solution:

Given

d =100 mm = 0.1m, length l =1m, m =1000 kg, E =200x10^9 N/m^2

Area of cross-section; $a = \frac{\pi d^2}{4} = \frac{\pi(0.1)^2}{4} = 7.854\ x10^{-3} m^2$

Moment of inertia of the shaft, $I = \frac{\pi d^4}{64} = \frac{\pi(0.1)^4}{64} = 4.9x10^{-6}\ m^4$

Frequency of longitudinal vibration:

Static deflection of shaft; $\delta = \frac{Fl}{AE} = \frac{1000x9.81x\,1}{7.854x10^{-3}x200x10^9} = 6.245x10^{-6}$ m

Natural frequency $f_n = \frac{1}{2\pi}\sqrt{\frac{g}{\delta}} = \frac{1}{2\pi}\sqrt{\frac{9.81}{6.245x10^{-6}}} = 199.5\ Hz$

Frequency of Transverse vibration:

Static deflection of shaft; $\delta = \frac{Wl^3}{3EI} = \frac{1000x9.81x\,(0.1)^3}{3x200x10^9x4.91x10^{-6}}$ =3.33 x10^{-3} m

Natural frequency $f_n = \frac{1}{2\pi}\sqrt{\frac{g}{\delta}} = \frac{1}{2\pi}\sqrt{\frac{9.81}{3.33x10^{-3}}} = 8.638\ Hz$

Example 5: Determine the natural frequency of the spring systems shown in Fig. 5.12. System carries of mass of 25 kg and spring stiffnesses K_1=10 kN/m and K_2=15 kN/m.

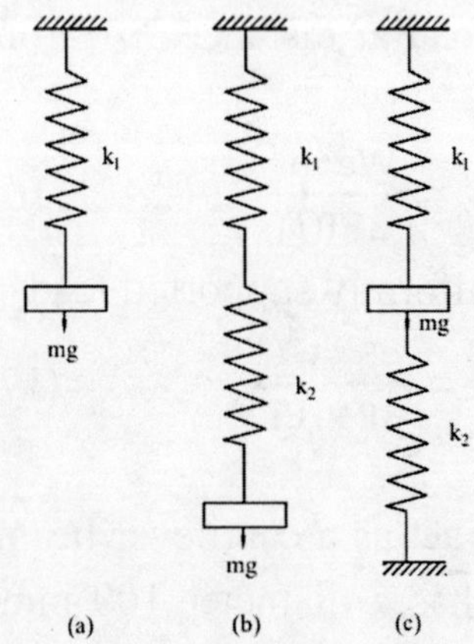

5.12 System of Springs; Ex. No. 5.

Solution:

Case a: Single spring with a mass

The equivalent spring stiffness k = 10 kN/m;

$$\text{Natural frequency}= f_n = \frac{1}{2\pi}\sqrt{\frac{\text{k}}{\text{m}}}$$

$$= \frac{1}{2\pi}\sqrt{\frac{10\text{x}10^3}{25}} = 3.183 \text{ Hz.}$$

Case b: Two springs in series with a mass

The equivalent spring stiffness $\text{k} = \text{k}=\frac{k_1k_2}{k_1+k_2} = \frac{10x15}{10+15} = 6$ kN/m;

$$\text{Natural frequency}= f_n = \frac{1}{2\pi}\sqrt{\frac{\text{k}}{\text{m}}}$$

$$= \frac{1}{2\pi}\sqrt{\frac{6\text{x}10^3}{25}} = 2.46 \text{ Hz.}$$

Case c: Two springs in series with a mass at centre fixed between rigid ends.

The equivalent spring stiffness $\text{k} = \text{k}_1+\text{k}_2 = 10+15 = 25$ kN/m;

$$\text{Natural frequency}= f_n = \frac{1}{2\pi}\sqrt{\frac{\text{k}}{\text{m}}}$$

$$= \frac{1}{2\pi}\sqrt{\frac{25\text{x}10^3}{25}} = 5.03 \text{ Hz.}$$

Example 6: A shaft of diameter 50 mm and length 1.5 m is simply supported as shown in Fig. 5.13; it carries a mass of 100 kg at 0.5 m from the left end. Find the natural frequency of transverse vibration. Assume E = 200 GN/m^2.

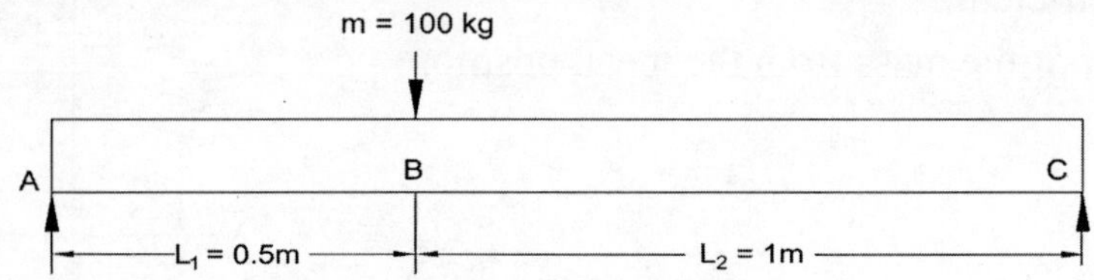

Fig.5.13 Shaft as a beam Ex. No. 6.

Solution:

Given

d =50mm = 0.05m, m =100 kg, E =200x10^9 N/m^2, a = 0.5 m, b =1.0 m,

Moment of inertia of the shaft, $\text{I} = \frac{\pi \text{d}^4}{64} = \frac{\pi(0.05)^4}{64} = 0.307 \text{ x}10^{-6} \text{ m}^4$

Dcflcction at point B

$$\delta = \frac{Wa^2b^2}{3EI(l)} = \frac{(100x9.81)(0.5)^2(1.0)^2}{3x\ 200x10^9x\ 0.307\ x10^{-6}\ x(1.5)}$$

$$= 0.888x10^{-3}\ m$$

$$\text{Natural frequency} = f_n = \frac{1}{2\pi}\sqrt{\frac{1}{\delta}}$$

$$= \frac{1}{2\pi}\sqrt{\frac{1}{0.888x10^{-3}}} = 16.73\ Hz.$$

5.10 Frequency of Damped Free Vibrations

In nature, the motion of a body is usually resisted by frictional forces and the effect of friction is referred to as damping. This led to the gradual diminishing of the amplitude of the vibrations. The resistance to the motion is provided partly by the surroundings in which the vibration takes place and also by the internal friction. It is also provided by an element called a dashpot or other external damping devices.

Therefore, a vibrating system is represented by a spring and a dashpot attached to the mass as shown in Fig. 5.14. Or a mass is suspended with a spring and a dashpot (damper) is provided between the mass and the rigid support as shown.

Let

m = Mass of the system,

k = Stiffness of the spring,

c = Damping coefficient,

x = Displacement of the mass from the mean position.

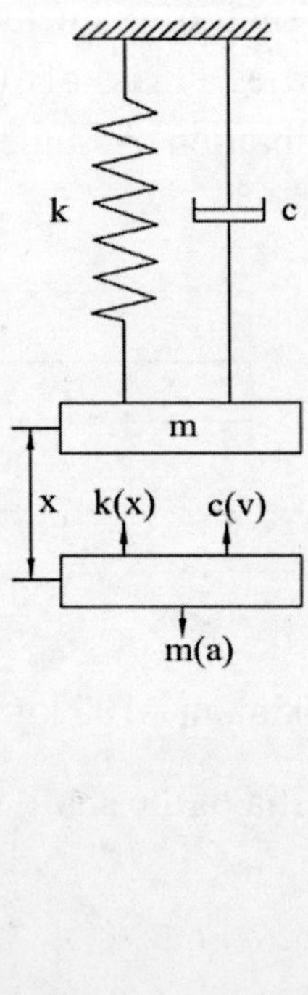

Fig.5.14 Spring Mass Damper System.

When the mass is displaced by a distance of x as shown in figure, the forces act on the mass are spring force equal to *(**k*** x distance), damping force (***c*** x Velocity) against the force (***m*** x acceleration).

The equilibrium equation can be written as

$$m\frac{dx^2}{dt^2} + c\frac{dx}{dt} + kx = 0 \qquad (18)$$

The Displacement of the mass from the mean position at the time (t) is given by the expression, $x(t) = C_1\, e^{R1\,(t)} + C_2\, e^{R2\,(t)}$ (19)

where C_1 and C_2 are the arbitrary constants and are to be determined from the initial conditions of the motion of the mass. The roots R_1 and R_2 may be real, complex conjugate or equal. The response of the system, i.e. the displacement (amplitude) of mass with time x(t), depends on the damping provided in the system. Those are compared with the critical damping equal to $C_c = 2m\omega_n$. The ratio of the actual damping coefficient (c) provided in the system to the critical damping coefficient (C_c) is called as damping factor or damping ratio ($c/2m\omega_n$).

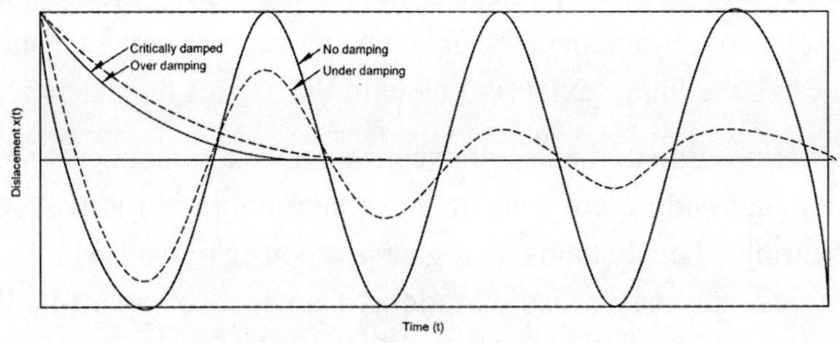

Fig.5.15 Motion of vibrating system for different damping coefficients.

The oscillation of the system or displacement of mass with time is shown in Fig.5.15 with

- No damping (c = 0): system vibrates with constant amplitude with time.
- Under damping ($c < C_c$): system vibrates with amplitude decay with time.
- Critical damping ($c = C_c$): motion is aperiodic and comes to equilibrium faster.
- Over damping ($c > C_c$): system motion is aperiodic and comes to equilibrium slowly

The damping coefficient, stiffness and mass of a system are carefully chosen or designed to achieve the purpose of the system.

5.11 Causes of vibrations in machines and remedies

Vibrations are developed due to one or more factors while machinery works at any given time. The most common reasons for vibrations are imbalance, misalignment, looseness, wear, and generally having improper lubrication.

- Imbalance: The vibrations will develop when the centre of mass of a rotating component is not aligned with the axis of rotation. The unbalanced weight rotates about the axis at higher speeds, creating a centrifugal force. These imbalances are caused by manufacturing defects, the use of components of different materials or maintenance issues. Balancing of car wheels when the new tyre is replaced, deposition of dirt chemicals on the turbine and missing bolts and nuts or balance weights are examples. The effects of imbalance become greater as the machine speed increases. The vibrations due to imbalance lead to unpleasant vibrations and comforts as well as a considerable reduction in the bearing life.

- Shaft runout: Vibration can result when machine shafts are out of line of axis. Angular misalignment occurs due to the reason that the axes of a drive and devices are not parallel or even exactly aligned. These conditions in the equipment are referred to as parallel misalignment. It can happen during the assembly or later with thermal expansion, component replacement, or improper maintenance. The resulting vibration can be radial or axial (in line with the axis of the machine) or both.

- Wear: Vibrations due to the use of excess worn-out components are very common. All rotating and sliding components have such vibration issues. Revolving shafts and the bearings, belts, chains, and gears in power transmission, sliding guides in machine tools, etc, may cause vibrations. Damaged components like pitted roller bearing races, gear teeth that are heavily chipped off, and peeled belts will cause a vibration each time they travel over the damaged area.

- Looseness: Machinery fitted with wrong-sized components or excessive wear of a component in an assembly leads to looseness among mating components and can develop vibrations. The looseness resulting in vibration can lead to further damages, such as wear and fatigue in equipment mounts and other complications.

Vibrations in machinery lead to many concerns: accelerated wear of components, consumption of excess power and demand for unplanned services. Increased repair and services or downtime are expensive, including loss of production. The other effects of vibration are creating an unpleasant working environment and safety issues.
The other useful side of vibration is that it indicates machine condition and calls for maintenance professionals to attend to the machinery and equipment service.
Remedies for vibrations involve the measurement of vibration signals from the machinery, particularly from the locations near the source of vibration. A transducer captures the required information and a diagnostic approach that plots individual vibration signals against frequency. This reveals the health and performance of the machinery.

The vibration analysis of machinery involves recognising the vibration peaks associated with specific components of the machine. Analysis of the data to identify patterns is very important in understanding the nature of vibration. Assessing the amplitude of the vibration peaks helps to estimate the severity of any detected fault.

Vibration reduces the seal's life and, therefore, the use of a dynamically balanced rotating assembly is preferred. Vibrations due to Misalignment between the driver and any machinery cause of vibration can be assessed by measuring both radial and axial components present. The use of flexible couplings reduces this type of vibration.

Unit Summary

- Balancing is correcting or eliminating the unwanted dynamic forces and moments in rotating machinery by adding or removing masses for the parts.
- The centrifugal force exerted by rotating mass m on the shaft is $F_c = m\omega^2 r$. This centrifugal force acting radially outward produces a bending effect on the shaft.
- A balancing mass m_b is attached in the same plane of rotation at a distance of r_b to balance this unwanted disturbing force.
- Several masses rotating in a plane is balanced by mass at a given radius; it is given by

$$m_b r_b = \sqrt{\left(-\sum (mr(\cos\theta))^2\right) + \left(-\sum mr(\sin\theta)\right)^2}$$

- And the position of the balancing mass at angle

$$\tan\theta_b = \frac{-\sum mr(\sin\theta)}{-\sum mr(\cos\theta)}$$

- Vibrations are the mechanical oscillations of any object about its equilibrium situation.
- When an elastic body is disturbed from its equilibrium position, a kind of motion is set up. It may be oscillatory and disturbing motion from its mean position. Then the body is said to be under vibration.
- After giving an initial displacement the body continues to vibrate without any external force acting on the body. Then, the body is said to be under free or natural vibrations.
- The natural frequency of a body is given by $\omega_n = \sqrt{\frac{k}{m}}$
- The most common reasons for vibrations are imbalance, misalignment, looseness, wear, and generally having improper lubrication.

Multiple Choice Questions

1. A mass m_1 attached to a rotating shaft at radius r_1 causes a disturbance. This can be reduced by adding a single mass m_2 attached at a radius in the same plane of rotation. Then, the correct relationship is given by
   (a) $m_1.r_2 = m_2.r_1$ (b) $m_1.r_2 = m_2.r_1$ (c) $m_1.m_2 = r_1.r_2$ (d) $m_1.r_1 = m_2.r_2$
2. The centrifugal force due to a disturbing mass m located at radius r on the shaft, running a uniform speed ω is given by
   (a) $m\omega r$ (b) mr (c) $m\omega^2 r$ (d) $m\omega r^2$
3. The amplitude of every cycle of vibration reduces systematically. Then such a body is said to be in
   (a) undamped vibration (b) forced vibration (c) damped vibration (d) free vibration
4. A body experiences transverse vibrations, and then the stresses induced in the body is
   (a) shear stress (b) bending stress (c) compressive stress (d) tensile stress
5. The particles of a body under longitudinal vibrations move
   (a) parallel to its axis (b) perpendicular to its axis (c) circularly about its axis (d) elliptically about its axis.
6. A particle executing with SHM has the time period T_p
   (a) $\pi/2\omega$ (b) $2\pi/\omega$ (c) $\sqrt{2}\pi/\omega$ (d) $\sqrt{\pi}/2\omega$
7. The damping coefficient of the vibrating body is equal to the critical damping. Then, the vibration of the body is
   (a) Oscillatory with constant amplitude (b) Oscillatory with reduced amplitude (c) aperiodic (d) Oscillatory with increased amplitude
8. A cantilever beam under vibration has a natural frequency ω_1 for the beam length L. If the beam is reduced to (L/2), then the natural frequency ω_2. The ratio of these natural frequencies (ω_2/ω_1) is
   (a) 1.0 (b) 2.0 (c) $2\sqrt{2}$ (d) $1/\sqrt{2}$
9. An overdamped vibrating system has a motion that reaches an equilibrium position.
   (a) faster than critically damped (b) same as critically damped (c) slower than critically damped (d) infinitely.
10. If the roots of the equation of motion for a spring-damper-mass vibrating body are real, then the system is
    (a) un damped (b) overdamped (c) under damped (d) critically damped

Answers to Multiple Choice Questions

1. (d), 2. (c), 3. (c), 4 (b), 5. (a), 6 (b), 7(c), 8. (c), 9(a), 10 (b)

Exercises

1. Four masses are attached to a shaft in the same plane. The magnitude of these masses A, B, C and D are 200 kg, 300 kg, 400 kg and 200 kg respectively. Their radii of rotation are 80 mm, 70 mm, 60 mm and 80 mm from the centre of rotation. The angular position of masses are A to B 45°, B to C 90° and C to D 120°. If the balancing mass rotates at a radius of 50 mm, find the magnitudes and angular position of balancing mass. [381 kg, 300°]

2. Four masses, P, Q, R and S, are 210 kg, 350 kg, 240 kg and 260 kg, respectively. Their radii of rotation are 0.25 m, 0.10 m, 0.20 m and 0.35 m respectively. The angular gap between the successive masses is noted to be 60°, 45° and 110°. Find the magnitude and position of the required to balance system, while the radius of rotation for the new mass is 0.2 m. [148.5, 124°]

3. A disk of mass 200 kg is revolving at 500 rpm. The centre of disk A is shifted to 25 mm from the centre of rotation. The disk is balanced by adding two masses at 145° and 260° from the line of reference of through point A. Find the magnitude of masses at radii 10 mm and 15 mm.

4. A shaft carries a circular disc and three masses, A, B and C attached 4 kg, 3 kg and 2.5 kg at radial distances of 75 mm, 85 mm and 50 mm. The angular positions of these masses are 45°, 135° and 240° respectively. Determine the counter mass for the static balance at a radial distance of 75 mm. [3.81 kg at 276º]

5. A shaft of diameter 75 mm and 500 mm long carries a disc of mass 250 kg at its free end. The Young's modulus for the shaft material is 200 GPa. Determine the frequency of longitudinal and transverse vibrations of the shaft.

6. A steel section of size 25 mm wide and 50 mm depth is simply supported at the endpoints 1 m apart. It carries a mass of 200 kg in the middle of the beam. Find the

frequency of transverse vibration for when the mass of the beam is neglected. Assume E = 200 GN/m^2. [5.5]

7. A simply supported hollow shaft 1.5 m long is supported at the ends and carries a wheel each of 50 kg mass at mid of the shaft. The shaft has an external diameter of 75 mm and an inner diameter of 37.5 mm. The density of the shaft material is 7500 kg/m^3. The Young's modulus for the shaft material is 200 GN/m^2. Calculate the frequency of transverse vibration of the shaft.

8. A weight of 25kgf is suspended from a rod of diameter 25mm and is 2 m long. Find the frequency of the system by neglecting the mass of the rod. The Young's modulus for the rod material is 200 GN/m^2 [720 Hz]

9. A spring-mass system has a natural frequency of ω1, and a second spring is added to this system so that the natural frequency ω_2 is half of ω_1. Determine the stiffness of the second spring. [$k_2 = k_1/3$]

10. A spring-mass system has a natural frequency of 10 Hz with a mass of 10kg. When a mass is altered, then the natural frequency of the second system is raised to 20 Hz. Calculate the mass of the second system. [2.5 kg]

Experiment

To balance the rotating disturbing masses with a balancing mass rotating in the same plane. (using Balancing apparatus, weights, nuts etc.)

Procedure:

- Note the masses and their location, i.e. radii and angular position.
- Fix these masses on a disk on the apparatus and confirm that they are firmly fixed.
- Turn ON the motor and observe the vibration of the rotating shaft.
- Determine the balancing mass, its radius and angular position by graphical and analytical methods.

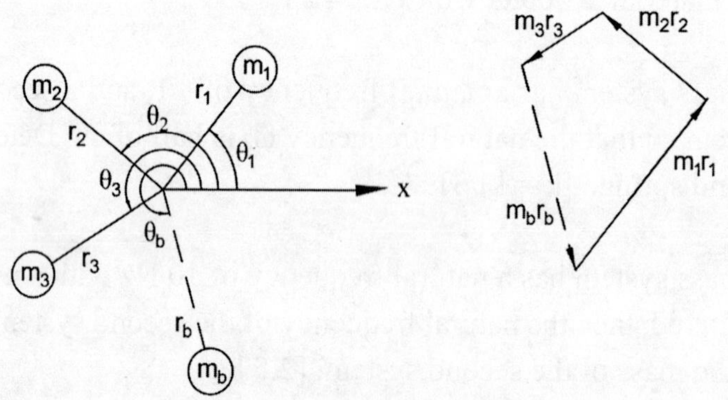

Fig.5.16 Balancing of masses in a plane.

Table 5.5 Resolved centrifugal forces.

Mass No.	Mass m (kg)	Radius r (m)	F = m.r (kg.m)	Angular Position (θ)	(mr)cosθ (kg.m)	(mr)sinθ (kg.m)

- Analytical method by using equations:

$$m_b r_b = \sqrt{\left(-\Sigma\left(mr(\cos\theta)\right)^2\right) + (-\sum mr(\sin\theta))^2}$$

$$\text{and} \quad \tan\theta_b = \frac{-\sum mr(\sin\theta)}{-\sum mr(\cos\theta)}$$

- The balancing mass and the disturbing were mounted at required radius and the respective calculated angles properly.
- Finally, the motor was turned on so that to observe the stability of the apparatus.
- Repeat experiment for different number masses, radii and angular positions.

KNOW MORE

Lecture Series on Kinematics of Machines by Prof. Asok Kumar Mallik, Department of Mech. Engg. IIT Kanpur. Watch the NPTEL video on YouTube using the links:
https://www.youtube.com/watch?v=HKVvJWArgg8
https://www.youtube.com/watch?v=hWNpID0TWYU

Bibliography

- Theory of Machines, RS Khurmi and JK Gupta, S. Chand Publishing, 2005.
- Theory of Machines, S. S. Rattan, McGraw Hill, 4th Edition, 2019.
- Theory of Machines and Mechanisms, John J. Uicker et al., Oxford University Press, Fifth Edition, 2017.
- Theory of Mechanisms and Machines, Amitabha Ghosh and Asok Kumar Mallik, East-West Press Private Limited, 1998.
- Theory Mechanisms and Machine, Jagdish Lal, Metropolitan Book Pvt Ltd., 1994.
- Mechanical Vibrations, S.S. Rao, Prentice Hall, 2011.
- https://www.youtube.com/watch?v=HKVvJWArgg8
- https://www.youtube.com/watch?v=hWNpID0TWYU

References and suggested readings

1. Theory of machines – S.S. Rattan, Tata McGraw-Hill publications, 5th ed., 2019.
2. Theory of machines – R.K. Bansal, Laxmi publications, 4th ed., 2016.
3. Theory of machines – R.S. Khurmi & J.K.Gupta, S.Chand Publications, 2005.
4. Dynamics of Machines – J B K Das, Sapna Publications, 2008.
5. Theory of machines – Jagdishlal, Bombay Metropolitan Book Ltd., 1985.
6. Theory of Machines – Thomas Bevan, Pearson Education 3rd Ed., 2010.
7. Theory of Machines – Sadhu Singh, Pearson Education, 2013.
8. Mechanism and Machine Theory – J. S. Rao, Rao V. Dukkipati, New Age Int, 2007.

CO AND PO ATTAINMENT TABLE

Course outcomes (COs) for this course can be mapped with the programme outcomes (POs) after the completion of the course and a correlation can be made for the attainment of POs to analyze the gap. After proper analysis of the gap in the attainment of POs necessary measures can be taken to overcome the gaps.

Table for CO and PO attainment

Course Outcomes	Expected Mapping with Programme Outcomes (1- Weak Correlation; 2- Medium correlation; 3- Strong Correlation)						
	PO-1	PO-2	PO-3	PO-4	PO-5	PO-6	PO-7
CO-1							
CO-2							
CO-3							
CO-4							
CO-5							

INDEX